T0135871

Applying Robust Scale M-Estimators to Compute Credibility Premiums in the Large Claim Case

Vom Fachbereich Mathematik
der Technischen Universität Darmstadt
zur Erlangung des Grades eines
Doktors der Naturwissenschaften
(Dr. rer. nat.)

genehmigte

Dissertation

von
Dipl.-Math. Annett Keller
aus Dresden

Referent:	Prof. Dr. J. Lehn
Korreferenten	Prof. Dr. A. May
	Prof. Dr. M. Kohler
Tag der Einreichung:	27. Februar 2008
Tag der mündlichen Prüfung:	27. Mai 2008

Darmstadt 2008
D 17

Bibliografische Information der Deutschen Nationalbibliothek

Die Deutsche Nationalbibliothek verzeichnet diese Publikation in der
Deutschen Nationalbibliografie; detaillierte bibliografische Daten sind
im Internet über http://dnb.d-nb.de abrufbar.

ISBN 978-3-8325-2037-3

Logos Verlag Berlin GmbH
Comeniushof, Gubener Str. 47,
10243 Berlin
Tel.: +49 030 42 85 10 90
Fax: +49 030 42 85 10 92
INTERNET: http://www.logos-verlag.de

Acknowledgements

This thesis has been developed during my employment as research assistant at Technische Universität Darmstadt.

First of all, I want to thank my supervisor Professor Jürgen Lehn for his support. I also would like to thank Professor Angelika May for being referee of my thesis. Especially, I am deeply grateful for her steady encouragement and advice. Our discussions have been a great inspiration for me.
Furthermore, I am grateful to Professor Michael Kohler for his willingness to be co-referee of this thesis and I would like to thank Professor Mirjam Dür and Professor Reinhard Farwig for being examiners in my dissertation committee.

I also want to thank Professor Jakob Creutzig, Dr. Katrin Schumacher and Dr. Andreas Rößler for their time and patience to answer my questions. Special thanks goes to Dominique who has been a great moral support.
Further thanks goes to my colleagues of the Stochastic Research Group for the pleasant working atmosphere.

Last but not least I want to thank my family and all my friends. That work would not have been possible without their never-ending support and faith in me.

Contents

List of Figures

List of Tables

Introduction

An *insurance premium* is the monetary amount the *insured* pays the *insurer* – usually an *insurance company* – for covering a specified *risk*.
An important branch in insurance mathematics is the pricing of possible large claims that are either the results of many small claims occuring at once or that are caused by single events. The underwriting of such claims is usually done by so called *re-insurance companies*.
As examples for literature that deals with this subject, we name here the works of Borch [Borch 1960] and [Borch 1962], Gerber [Gerber 1977], d'Ursel, Lauwers [d'Ursel, Lauwers 1985], Denuit et al. [Denuit et al. 2001], Schmidli [Schmidli 2001] and Ladoucette, Teugels [Ladoucette, Teugels 2006] that all illuminate different aspects of re-insurance premium calculation.
The ongoing interest in this subject is of no surprise since industrial insurances against fire loss and (re-)insurances against losses due to natural disasters face a remarkable increase in the claim amounts, cf. sigma-report [sigma 2003].

Even though we are able to give an explicit definition of an insurance premium, it is not that easy with the term *risk*.
In general, literature distinguishes between risk in a financial and in an insurance context. For the latter, usually a non-negative random variable is employed describing the claim amount caused by the insured, cf. Bühlmann, Straub [Bühlmann, Straub 1970]. The book of Rolski et al. [Rolski et al. 1999], which can be regarded as a standard book in insurance mathematics, refers to this definition as well, saying "any non-negative random variable or its distribution function" is called a risk. Borch [Borch 1974] improves the definition by introducing the reserve parameter R, a monetary amount the insurance company can use to settle claims.
The broadest definition of insurance risk is given by Mack [Mack 1997], another standard book, saying "risk is the smallest unit that can be insured".

When talking about financial risk, a non-negative random variable represents the asset or stock price assigned to the risk, cf. Cox et al. [Cox et al. 1979]. As risk in an insurance context stands for possible future losses, risk from a financial point of view is seen as "the uncertainty of future earnings" by Martin et al. [Martin et al. 2006], p.1.
Lately, national and international regulatory authorities ask both insurance and banking companies to establish surveillance tools for the identification and quantification of risks inhered in their portfolios. Basel II and Solvency II standards, [Basel 2006],

[Solvency] lead to research in these areas, cf. Sandström [Sandström 2007] and Vesa et al. [Vesa et al. 2007] for example.

Today, actuaries have different means to calculate an insurance premium such as the expected value principle, the variance principle or the principle of equivalent utility along with the Swiss premium principle. All of them are widely explained and discussed in standard actuarial literature, e.g. Bühlmann [Bühlmann 1970], Goovaerts et al. [Goovaerts et al. 1984], Bowers et al. [Bowers et al. 1997] and Rolski et al.[Rolski et al. 1999]. Early work, regarding premium principles using a utility function, was done by Gerber [Gerber 1974], Bühlmann and Jewell [Bühlmann, Jewell 1979], [Bühlmann 1980]. Research has been continued for example by Browne [Browne 1995], Chateauneuf [Chateauneuf 1999] and Luan [Luan 2001]. Good references for premium calculation are also Sundt [Sundt 1999], who gives a comprehensive introduction on the main topics in non-life insurance and Mikosch [Mikosch 2004], who approaches the subject employing stochastic processes.

Because the occurence of single events causing such large claim amounts, is seldom, methods of extreme value theory found their ways into reinsurance mathematics, cf. Embrechts, Schmidli [Embrechts, Schmidli 1994], Asmussen [Asmussen, Klüppelberg 1996], Embrechts et al. [Embrechts et al. 2003] and again Mikosch [Mikosch 2004]. Lately, also the transfer from insurance risk into the financial market has been discussed, for example in Jaffee, Russel [Jaffee, Russel 1996], Herlihy, Parisi [Herlihy, Parisi 1999] and Christensen, Schmidli [Christensen, Schmidli 2000], Munich Re [Munich Re 2001], as well as Dassios, Jang [Dassios, Jang 2003].

A premium calculation principle that emphasises the structure of an insurance portfolio is the so called *credibility premium*. That method has first been discussed by Bühlmann [Bühlmann 1967] and [Bühlmann 1970] as well as Jewell [Jewell 1974]. Further research has been done by Marazzi [Marazzi 1976], Dubey, Gisler [Dubey, Gisler 1981], Eichenauer et al. [Eichenauer et al. 1988] and Schmidt [Schmidt 1990], Künsch [Künsch 1992], Gisler, Reinhard [Gisler, Reinhard 1993], Dannenburg [Dannenburg 1996]. Recent works are from Goulet [Goulet 2001], Purcaru, Denuit [Purcaru, Denuit 2002] and Pan et al. [Pan et al. 2008]. The books of Sundt [Sundt 1999], Herzog [Herzog 1999] and Bühlmann, Gisler [Bühlmann, Gisler 2005] give a good introduction and summary to credibility theory.

The credibility premium is a convex combination of the *class mean*, representing the insurance portfolio's general behaviour and the *individual mean*. The latter takes into account the individual claim history of the risks subsumed in the portfolio, cf. Bühlmann [Bühlmann 1970] and Gerber [Gerber 1979]. The insurer calculating the premium does not necessarily need to know the claim amount distribution, even though she has to make some assumptions, for example the existence of first and second moments.

In this thesis an insurance portfolio of N risks – then called *risk classes* – is considered.

It is assumed that each of the risks typically causes a small claim amount during an insurance period. But once in a while, the risks may produce large claim amounts due to a contamination of the small claim amount distribution function.

Because the classical credibility premium relies on the arithmetic mean, such rare but large claim amounts may cause problems. Since the mean is very sensitive towards outliers, large claims within a sample of mainly small claims will lead to an overestimation of the individual mean. The result is a credibility premium not reflecting the actual claim amount distribution in the portfolio.

For such models to calculate an insurance premium, the credibility approach can be applied combined with methods from *robust statistics*, cf. Künsch [Künsch 1992], Gisler, Reinhard [Gisler, Reinhard 1993]. We will follow their approaches separating both the claim amounts and the insurance premiums into *ordinary* and *extreme* parts.

The premium for the ordinary part is determined by applying the credibility principle. We assume that the claim amount distribution function F_{θ_i} of risk i, $i = 1, \ldots, N$ is parametrised by a risk parameter θ_i, being a random variable itself. The distribution function of the independent risk parameters θ_i is known and called *structure function* since it induces a certain behaviour in the portfolio considering the claim amounts.
The parametrisation of the risk's claim amount distribution function is such that the expected value conditioned on θ_i is either equal to the parameter itself or some linear transformation of it.

The rare, large claim amounts originate from a contamination of the claim amount distribution function F_{θ_i}. Thus, instead of applying the arithmetic mean as an estimator for θ_i – as it is done in classical credibility theory – we will introduce a robust estimator $T_{n,i}$. By identifying and neglecting large claims within each risk class i, the estimator $T_{n,i}$ gives a much better prediction of the risk parameter θ_i. Of course, then the calculation of the *credibility factor* has to be adjusted as well.
Determining the premium of the extreme part, the mean excess function is going to be used.

The thesis is organised as follows.
We start in Chapter 1 with surveys of insurance premiums and robust statistics. Furthermore, we define the data model that is going to be applied in this thesis.

In Chapter 2, the view is more focused on quantitative statistics. First, we introduce the class of so called M-estimators, a general form of maximum likelihood estimators.
For a random sample $X_1, \ldots, X_n$, the M-estimator θ_M is defined by

$$\sum_{j=1}^{n} \psi(X_j; \theta_M) = 0,$$

where ψ is some non-decreasing function. We will discuss that ψ chosen to be

$$\psi(x; \theta) = \max\left\{-b, \min\left\{\frac{\partial}{\partial \theta} \ln f_\theta(x) - a, b\right\}\right\}, \qquad f_\theta(x) = \frac{d}{dx} F_\theta(x),\ a \in \mathbb{R},\ b > 0$$

is optimal in the sense that θ_M is of minimal variance.

Additionally, more concepts of quantitative robustness are presented with a special emphasis on *influence functions*. It is also shown how influence functions can be used to derive robust M-estimators.

The main results of this thesis are presented in Chapter 3. Based on the function ψ as given above, we will define a robust M-estimator T_n with respect to our data model. The estimator T_n is going to be consistent. But its calculation rule is somewhat uncomfortable because it does depend on T_n itself. By slightly modifying the function ψ yielding $\tilde{\psi}$, we will define the robust estimator $\tilde{T}_n$ which does not have the computational problems of $\tilde{T}_n$.

We also discuss the question of choosing the parameter a for ψ and $\tilde{\psi}$ in an optimal way.

In Chapter 4 we are going to compute some of the quantitative characteristics introduced in Chapter 2 for both T_n and $\tilde{T}_n$, namely the gross errors and the finite sample breakdown points. We also prove that both estimators T_n and $\tilde{T}_n$ are asymptotic normally distributed.

The thesis is completed by a simulation study in Chapter 5. We analyse the sensitivity of T_n and $\tilde{T}_n$ towards different choices of a and b, as well as changing sample sizes and possible occurings of large claims. The simulation will show that for reasonable choices of a and b, the robust estimators T_n and $\tilde{T}_n$ can bear the comparison with the median, which is known as the most robust estimator.

As well, we will estimate the credibility premiums for an insurance portfolio consisting of 25 risk classes and discuss the circumstances, when an actuary should apply the robust credibility approach.

Chapter 1

Preliminaries

1.1 The Term Insurance Premium

An insurance company usually sells insurance contracts against different random events causing possible losses to the insured, e.g. car liability insurance against car accidents or disability insurance in case the insured cannot continue working.
The risks an insurance company underwrites are grouped together in *portfolios* or *collectives* of similar risks. For example, car drivers living in the same town are grouped together, or employees with the same kind of job. Sometimes, the separation goes even further, using criteria such as age, gender or experience.

Insurance mathematics provides adequate means to model the situation of such a portfolio. An important role plays the *collective model*. In contrary to the *individual model*, within the collective model the focus is only on the overall claim amount X caused by the insured, cf. Schmidt [Schmidt 2006].
The number of such claims occuring in an insurance period within a portfolio is a random variable M. Thus the claim amount of the portfolio can be described by the random variable

$$S = \sum_{l=1}^{M} X_l, \tag{1.1}$$

a so called *compound sum process*, cf. Bowers et al. [Bowers et al. 1997].

Before we have a closer look at the characteristics and the determination of a *credibility premium*, we spend some thoughts on the term *insurance premium* in general.

A more detailed and extended explanation of the terms described below can be found in the books of Mack [Mack 1997], Bühlmann, Gisler [Bühlmann, Gisler 2005] or Schmidt [Schmidt 2006], for example.

The amount an insured pays his insurance company is called *gross premium*. It consists of the *pure risk premium* and some bonus covering safety loading, administrative costs and

the profit margin.
The pure risk premium indicates the amount of money used to cover the claim amount
X. Now, since the insurer never knows which claim amounts she faces during the next
insurance period, she uses the expected claim amount $E(S)$ as a prediction. If the single
claim amounts $(X_l)_{l \in \mathbb{N}}$ are independent and identically distributed (i.i.d.) then because of

$$E(S) = E(M) \cdot E(X),$$

the insurer knows the pure risk premium $E(X)$ for each insured as well. More on this
topic can be found in the books of Bowers et al. [Bowers et al. 1997] and Schmidt
[Schmidt 2006]. We call $E(S) = \mu_{\text{net}}$ the net premium.

Theory shows that charging only the net premium, the insurer will suffer a ruin event
almost surely, as it is explained by Mack [Mack 1997] and Schmidt [Schmidt 2006]. Thus
a *safety loading* has to be added to the net premium. The sum is then called *risk premium*.
Classical premium calculation principles such as the expected value principle, the variance
principle or the standard deviation principle take the safety loading into account.
Applying one of these calculation methods, the insurer gets a risk premium that can be
divided into a net premium and a safety loading.

This thesis deals with a premium calculation method other than the classical ones mentioned above.
We do neither touch the question of an adequate safety loading nor are we interested in the
amount of other costs or margins. The solely focus lies on determining the net premium.
Thus, from now on we denote the net premium briefly by *premium* or *insurance premium*.

1.2 Credibility Premium

As already mentioned in the introductory chapter, a credibility premium does not depend
on the exact distribution the data come from.
A comprehensive introduction to credibility theory can be found in the books of Herzog
[Herzog 1999], Sundt [Sundt 1999] as well as Bühlmann, Gisler [Bühlmann, Gisler 2005].
In the literature we find different types of credibility, e.g. limited fluctuation credibility,
greatest accuracy credibility, as defined by Jewell [Jewell 1976], or Bayesian credibility
mentioned in the works of both Herzog [Herzog 1999] and Sundt [Sundt 1999]. We only
have a closer look on the greatest accuracy credibility, because it leads to our model.

For Sections 1.2.1 and 1.2.2 we assume that all necessary probability spaces are defined in
an appropriate way.

1.2.1 A Short Survey of Credibility

The idea for calculating an insurance premium in the way described below goes back as
far as the beginning of the 20th century and can for example be found in the paper of
Whitney [Whitney 1918].

As it has already been pointed out, an insurer groups similar risks into collectives. Within these collectives, the risks are still heterogeneous enough to be categorised further into different risk classes $i = 1, \ldots, N$.

We denote the claim amounts X_{ij}, $i = 1, \ldots, N$, $j = 1, \ldots, n_i$ where N is the number of risks and n_i is the number of claims of risk i, we have within each risk class. Usually, these claim amounts are called the *experience* of risk class i. Therefore this approach is also called *experience rating*.

In the context of the collective model from Section 1.1, the overall claim amount S is

$$S = \sum_{i=1}^{N} \sum_{j=1}^{n_i} X_{ij}.$$

Note that in credibility theory we use data from previous insurance periods to derive a premium for the next period. Thus the number of claim amounts M in (1.1) is not a random variable any more. In fact, $M = \sum_{i=1}^{N} n_i$.

The insured's net premium $\mu_{\text{net}} = E(X)$ is called *collective premium* μ_{coll} in the credibility model. Whereas the net premium of a claim amount assigned to a certain risk class i is called *individual premium* μ_{ind}. Obviously both the insurer and the insured are interested in μ_{ind}.

Whitney suggests in [Whitney 1918] to determine the individual premium as a weighted sum of monetary amounts covering both the individual risk experience as well as the collective experience, which he calls *class experience*. He points out that the weigthing or *credibility factor* c should be closely related to the data.

In detail, this means that a large number of observations within risk class i, varying only little, should give the actuary confidence in the individual experience, thus shifting more weight to this side. On the other hand, having only few data, the actuary has to rely more on the class experience.

As we will see in the next section, Bühlmann improved the idea by providing the actuary with a general credibility model, cf. [Bühlmann 1967]. Since then a lot of research has been done, focusing on the credibility factor, for example by Dubey, Gisler [Dubey, Gisler 1981], Eichenauer et al. [Eichenauer et al. 1988] as well as Künsch [Künsch 1992], Gisler, Reinhard [Gisler, Reinhard 1993] and Pan et al. [Pan et al. 2008]. Generalisations of the original assumptions can be found in the works of Bühlmann, Straub [Bühlmann, Straub 1970] and Marazzi [Marazzi 1976]. Convergence results and credibility in continuous time has been treated by Schmidt [Schmidt 1990] and Merz [Merz 2004]. In what follows, we concentrate on the work of Bühlmann.

1.2.2 Credibility in the sense of Bühlmann

The development of the greatest accuracy credibility is closely connected with the name Bühlmann and his works [Bühlmann 1967] and [Bühlmann 1970].

He considers an insurance portfolio with risk classes $i = 1, \ldots, N$, as described above.

But in a step of enhancement, he assumes each risk class to be characterised further by a risk parameter ϑ_i, $i = 1, \ldots, N$, meaning that the claim amount distribution function for risk class i depends on ϑ_i.
Until then, the risk parameter has been regarded as something given exogenously.
Bühlmann looked at the risk parameter as a random variable, thus following a *Bayesian approach*, cf. [Bühlmann 1967], [Bühlmann 1970] and [Bühlmann, Gisler 2005].

The random variables $\vartheta_1, \ldots, \vartheta_N$ are assumed to be i.i.d. with values in a set $\Theta \subseteq \mathbb{R}$ and distribution function U. The distribution function is to be predefined by the actuary, accommodating her beliefs in the risk structure of the portfolio. Hence U is called *structure function*.
In Bühlmann's model, for any risk $i = 1, \ldots, N$, we have the same number $n_i = n$ of observations $X_{i1}, \ldots, X_{in}$ that are assumed to be i.i.d. for given $\vartheta_i \in \Theta$ with distribution function F_{ϑ_i}. Furthermore, the N vectors $(\vartheta_i, X_{i1}, \ldots, X_{in})$ are assumed to be i.i.d. for $i = 1, \ldots, N$.

The implementation of the risk parameter ϑ_i enables us to express the individual premium $\mu_{\mathrm{ind},i}$ of risk class i with respect to ϑ_i

$$\mu_{\mathrm{ind},i} = E(X_{i1}|\vartheta_i) =: \mu(\vartheta_i), \qquad i = 1, \ldots, N. \tag{1.2}$$

We note that $E(X_{i1}|\vartheta_i)$ is the net premium for risk class i.

Unfortunately, in practice neither ϑ_i nor $\mu(\vartheta_i)$ are known. We therefore call $\mu(\vartheta_i)$ the *hypothetical mean* as it has been done by Herzog [Herzog 1999] for example. The *variance of the hypothetical mean* $\mathrm{Var}(\mu(\vartheta_1))$ measures the heterogeneity among the different risk classes.
In contrast, the *process variance* $\sigma^2(\vartheta_i)$, $i = 1, \ldots, N$ indicates the deviation from the hypothetical mean within one risk class. The expected deviation is consequently called *expected value of the process variance* $E(\sigma^2(\vartheta_1))$. Again further explanations can be found in the books of Herzog [Herzog 1999] and Sundt [Sundt 1999].
Not knowing $\mu(\vartheta_i)$, we have to find an estimator $\widehat{\mu}(\vartheta_i)$.

It is also possible to express the collective premium in terms of the individual premium using the structure function

$$\mu_{\mathrm{coll}} = \int_{\Theta} \mu(\theta) dU(\theta).$$

Even though the derivation of the credibility premium does not demand the knowledge of the underlying claim distribution function, Bühlmann requires the existence of the first two moments of claim amounts. Since $X_{i1}, \ldots, X_{in}$ are i.i.d., this means $E(X_{i1}|\vartheta_i) = \ldots = E(X_{in}|\vartheta_i) =: \mu(\vartheta_i)$ and $\mathrm{Var}(X_{i1}|\vartheta_i) = \ldots = \mathrm{Var}(X_{in}|\vartheta_i) =: \sigma^2(\vartheta_i)$.

We already pointed out that by treating the risk parameters ϑ_i, $i = 1, \ldots, N$ as random variables, Bühlmann follows a Bayesian approach. Hence, it is not surprising that we can connect a Bayesian premium with a credibility premium.

At this point we only give a very short introduction to Bayesian analysis and refer to the standard books of Berger [Berger 1985] and Lehmann, Casella [Lehmann, Casella 1998] for detailed information.

In Bayesian analysis and decision theory, *prior distributions* and *loss functions* play an important role. The usual setup is that the outcome X of some random experiment depends on a parameter ϑ, itself regarded as a random variable. The distribution function π of ϑ, called prior distribution, is known but constantly adjusted to the random sample $\mathbf{X} = (X_1, \ldots, X_n)$. The adjusted distribution $\pi(\vartheta|\mathbf{X})$ is called *posterior distribution*, cf. Berger [Berger 1985].

Now estimating the random variable ϑ and approaching it from a Bayesian point of view, we get as the optimal Bayes estimator the *posterior mean* $\delta(\mathbf{X}) = E(\vartheta|\mathbf{X})$ provided that we apply a quadratic loss function, cf. Lehmann, Casella [Lehmann, Casella 1998].

Passing this in our context, the Bayes estimator is called the Bayes premium

$$\mu_{\text{Bayes}} = E(\mu(\vartheta_1)|\mathbf{X}),$$

cf. Bühlmann [Bühlmann 1967] and Bühlmann, Gisler [Bühlmann, Gisler 2005].

The credibility premium $\hat{\mu}_{\text{C},i}$ of risk class i is then the linear approximation of the Bayes premium. Detailed calculations proving this, has been done by Bühlmann [Bühlmann 1970] and Sundt [Sundt 1999], for example.

Here we only state the result

$$\hat{\mu}_{\text{C},i} := \widehat{\mu}(\vartheta_i) = c_i \cdot \bar{X}_i + (1 - c_i) \cdot \mu_{\text{coll}}, \qquad i = 1, \ldots, N \tag{1.3}$$

with

$$\bar{X}_i = \frac{1}{n} \sum_{j=1}^{n} X_{ij}, \qquad \mu_{\text{coll}} = E(\mu(\vartheta_1))$$

$$c_i = \frac{n \cdot \text{Var}(\mu(\vartheta_i))}{n \cdot \text{Var}(\mu(\vartheta_i)) + E(\sigma^2(\vartheta_i))},$$

cf. Bühlmann [Bühlmann 1967]. Reflecting the average claim amount of the portfolio, μ_{coll} is also called *class mean*.

Since we can actually determine $\bar{X}_i$ for each $i = 1, \ldots, N$, we call it *individual mean*. Because $\vartheta_1, \ldots, \vartheta_N$ are i.i.d., we have

$$c_i = c = \frac{n \cdot \text{Var}(\mu(\vartheta_1))}{n \cdot \text{Var}(\mu(\vartheta_1)) + E(\sigma^2(\vartheta_1))} = \frac{n}{n + \frac{E(\sigma^2(\vartheta_1))}{\text{Var}(\mu(\vartheta_1))}}. \tag{1.4}$$

It is easy to see that an increase in the variance $\text{Var}(\mu(\vartheta_1))$ of the hypothetical mean moves c closer to 1. This is an appropriate behaviour since $\text{Var}(\mu(\vartheta_1))$ explains the variation within the collective. Therefore more credibility should be put in the individual mean because of the collective's heterogeneity.

On the other hand, a big value of the expected process variance $E(\sigma^2(\vartheta_1))$ indicates wide spreading of the individual experiences around their means within the risk classes. It then is more reasonable to shift the weight in (1.3) towards the class mean μ_{coll}.

Having estimators for the unknown hypothetical means $\mu(\vartheta_i)$, $i = 1, \dots, N$, we are not really better off since we do not know μ_{coll}, $E(\sigma^2(\vartheta_1))$ and $\text{Var}(\mu(\vartheta_1))$. By again installing appropriate estimators $\hat{\mu}$, $\hat{E}(\sigma^2(\vartheta_1))$ and $\widehat{\text{Var}}(\mu(\vartheta_1))$, we get the *empirical credibility premium*

$$\hat{\mu}_{C,i;\,n} = \hat{c} \cdot \bar{X}_i + (1 - \hat{c}) \cdot \hat{\mu}, \qquad \hat{c} = \frac{n \cdot \widehat{\text{Var}}(\mu(\vartheta_1))}{n \cdot \widehat{\text{Var}}(\mu(\vartheta_1)) + \hat{E}(\sigma^2(\vartheta_1))}. \tag{1.5}$$

Note that in this simple Bühlmann model we assume the same number n of oberservations in each risk class $i = 1, \dots, N$.

For the reason of completeness we mention at this stage, that Bühlmann and Straub developed a credibility approach that can handle different numbers of observations. More information on this can be found in the paper of Gisler, Reinhard [Gisler, Reinhard 1993] and the book of Bühlmann, Gisler [Bühlmann, Gisler 2005].

1.3   A Short Survey of Robust Statistics

First, we are going to define the term *statistical model* in a general statistical setup, which later on we will use in a robust context. The definition bases on the books of Müller [Müller 1991] and Bernardo, Smith [Bernardo, Smith 2000].

Definition 1.1
Let $X_1, \dots, X_n$ be a random sample on a probability space $(\Omega, \mathfrak{A}, P)$ with values in a measure space $(\Omega', \mathfrak{A}')$. A statistical model $(\Omega', \mathfrak{A}', \mathbf{Q})$ is a set of distributions $\mathbf{Q}$ on $(\Omega', \mathfrak{A}')$. If the set $\mathbf{Q}$ can be parametrised, then the model is called parametric statistical model.

So far we have learned that the credibility premium is defined as a convex combination of the individual mean $\bar{X}_i$, $i = 1, \dots, N$ and the class mean μ_{coll}. Even though the credibility factor c is designed to cover up to a certain stage of heterogeneity, it cannot compensate for outliers.

Inferences in statistics mainly rely on the quality and quantity of the data and the assumptions the statistician bases her model on, such as the underlying model distribution or the independence of observations.

Unfortunately, even though these assumptions are chosen with great care, they might turn out to be wrong. Therefore, if the statistician lacks sufficient or reliable information to define an accurate model, it will be reasonable to apply estimators that do not only predict the assigned quantity as accurate as possible but compensate discrepancies in the assumptions or data up to a certain level as well. Such estimators are called *robust*. Loosely speaking, robust estimators cut off outliers.

Now the discrepancies in the assumptions are for example a misjudgement of independence of the observations or a wrong choice of the assumed underlying distribution. According to these different occurrences of deviations from the assumed statistical model, we distinguish between two types of robustness: *qualitative* and *quantitative robustness*.

Connected with the question of the model distribution is the derivation of classical estimators. Popular and well known techniques such as the least squares method or concepts as the squared error loss function rely on the second moments of the random variable. But many random variables with heavy-tailed distributions do not have second moments, some do not even have first moments, cf. Johnson et al. [Johnson et al. 1994].

1.3.1 Qualitative Robustness

Even though Huber [Huber 1964], was the first to introduce a systematic view on robust statistics, it was Hampel in [Hampel 1971] who gave a definition of what is now regarded as *qualitative robustness*. This enables us to study deviations from the true model with respect to the underlying distribution.

In Chapter 2 we will explain that an estimator can be regarded as a functional T defined on the set of distribution functions and mapping into $\mathbb{R}$. Therefore deviations from the true statistical model can be examined by applying metrics on distributions or distribution functions.
For example, Hampel [Hampel 1971] follows this approach.

Definition 1.2
Let (Ω, d) be a complete metric space and $(\Omega, \mathfrak{A})$ be a measurable space. For $A \in \mathfrak{A}$ and $\varepsilon > 0$ define

$$A^\varepsilon = \{\omega \in \Omega : \inf_{x \in A} d(x, \omega) < \varepsilon\}.$$

Let P, Q be two probability measures on $(\Omega, \mathfrak{A})$, then the Prohorov *distance d_{Proh} of P and Q is defined as*

$$d_{Proh}(P, Q) = \inf\{\varepsilon > 0 : P(A) \leq Q(A^\varepsilon) + \varepsilon, \forall A \in \mathfrak{A}\}.$$

Now Hampel [Hampel 1971] defines a *robust estimator* T_n or rather a *robust sequence of estimators* $(T_n)_{n \in \mathbb{N}}$ as in

Definition 1.3

A sequence of estimators $(T_n)_{n \in \mathbb{N}}$ *is called* robust *at a probability measure* P *if for all* $\varepsilon > 0$ *there exist* $\delta > 0$ *such that for all probability measures* $Q \in \mathcal{P}$ *and for all* $n \in \mathbb{N}$:

$$d_{Proh}(P, Q) < \delta \implies d_{Proh}(\mathfrak{L}_P(T_n), \mathfrak{L}_Q(T_n)) < \varepsilon,$$

where $\mathcal{P}$ *is the set of probability measures on* $(\Omega, \mathfrak{A})$.

In the above definition $\mathfrak{L}_P(T_n)$ and $\mathfrak{L}_Q(T_n)$ denote the distribution of the estimator T_n under the probability measures P and Q.

Shortly speaking, qualitative robustness deals with the problem of choosing a distribution close to the true one. In contrary, quantitative robustness focuses on the data itself.

1.3.2   Quantitative Robustness

Working in statistics always means to extract certain information from the data.
But we cannot only use the data to maintain estimates or perform hypothesis tests, we can also use the data to find out more about the estimator itself.
In most cases, this turns out to be easier than checking for qualitative robustness, since we do have the data but we do not always have the probability measures or distributions.

Quantitative robustness looks at an estimator from different perspectives. The *breakdown point* tells us, how much impact outliers have on the estimate. In comparison, the *empirical influence function* describes the impact of one observation. Its maximum value is called *maximum bias*.
Often there is also a special interest in the asymptotic behaviour of the estimator.

At the moment, we will not explore the field of quantitative robustness any further but give some references instead. The exact definition of the quantities listed above is postponed until we have defined an actual estimator.

In Hampel's work [Hampel 1974], the reader will find a detailed introduction to the application of the influence function in robust statistics, combined with historical notes on the topic. Thall [Thall 1979] discusses a robust scale estimator for the Exponential distribution that corresponds to the location estimator of Huber [Huber 1964].
The works of Rousseeuw, Leroy [Rousseeuw, Leroy 1988] and Gather, Schultze [Gather, Schultze 1999] deal with certain aspects of quantitative robustness. Recent research on robust scale estimators has been done by Collins for example [Collins 1999], [Collins 2003] and Szatmari, Collins [Szatmari, Collins 2007].
We also refer to the books of Hampel et al. [Hampel et al. 1986] and Jurečková, Picek [Jurečková, Picek 2006], both of them giving an overview of the knowledge in robust statistics so far.

1.4 The Model Setting

The last sections were devoted to the discussion of the credibility premium and a short overview of robust statistics. In this section we describe the statistical model we will work with in the following.

Let $(\Omega, \mathfrak{A}, P)$ be a probability space and let $X_{ij} : \Omega \to (0, \infty)$, $i = 1, \ldots, N$, $j = 1, \ldots, n$ be random variables on $(\Omega, \mathfrak{A}, P)$. We regard X_{ij} as the jth claim amount of risk class i. In this work we consider the Bühlmann model, described in [Bühlmann 1967] and [Bühlmann 1970], meaning that for all risk classes i there is the same number n of observations.

As it has been suggested by Bühlmann, we also associate each of the N risk classes with a risk parameter ϑ_i. The random variables $\vartheta_1, \ldots, \vartheta_N$ are i.i.d. with structure function U and values in $\Theta \subseteq \mathbb{R}$.

We assume that each risk i mainly yields claim amounts X_{ij} that have distribution function F_{ij,ϑ_i} parametrised by the risk parameter ϑ_i. Furthermore, given ϑ_i, the random variables $X_{i1}, \ldots, X_{in}$ are conditionally independent and identically distributed. Thus we actually have $F_{ij,\vartheta_i} = F_{\vartheta_i}$.

Seldom, risk i causes claim amounts that deviate widely from the remaining in the sense that they are much bigger.

This situation can be described applying the so called ε-*contamination model* $\mathcal{F}_\varepsilon$

$$\mathcal{F}_\varepsilon(F_{\vartheta_i}) = \{F \colon F = (1 - \varepsilon) \cdot F_{\vartheta_i} + \varepsilon \cdot G, \ G \in \mathcal{F}\}, \tag{1.6}$$

where $0 < \varepsilon < 1$, F_{ϑ_i} is given and G is some distribution function. We will choose G to be a heavy-tailed continuous distribution function. The above modelling is for example presented in the paper of Gisler, Reinhard [Gisler, Reinhard 1993]. More information on heavy-tailed distribution functions can be found in the book of Johnson et al. [Johnson et al. 1994].

Since we know that risk i causes extreme claim amounts, we split the individual premium $\mu_{\text{ind},i}$ covering risk i in a part $\mu_{\text{ord},i}$ associated with the non-extreme claim and a part μ_{xs} that is associated with the extreme claim part,

$$\mu_{\text{ind},i} = \mu_{\text{ord},i} + \mu_{\text{xe}}. \tag{1.7}$$

Gisler and Reinhard speak of an *ordinary*- and an *xs*-(excess-)part, [Gisler, Reinhard 1993]. We adapt their notation in this work.

Choosing this perception, we can calculate the premium $\mu_{\text{ord},i}$ as a credibility premium in the sense of Bühlmann, [Bühlmann 1967]. That means, the claim amount distribution F_{ϑ_i} for risk class i depends on a risk parameter ϑ_i, $i = 1, \ldots, N$. For the premium $\mu_{\text{ord},i}$ then it is reasonable to assume

$$\mu_{\text{ord},i} = \mu(\vartheta_i), \qquad i = 1, \ldots, N,$$

cf. equation (1.2).

As it is suggested by Bühlmann, the estimator $\hat{\mu}_{C,i}(\vartheta_i)$ is a convex combination of two different means, cf. equation (1.3). But now, we will not use the individual mean $\bar{X}_i$ because some of the claim amounts X_{ij} contain extreme parts. Instead, we install a robust estimator $T_{n,i}$.
Thus our robust credibility premium for $\mu(\vartheta_i)$ is

$$\hat{\mu}_{rC,i}(\vartheta_i) = c_{r,i} \cdot T_{n,i} + (1 - c_{r,i}) \cdot \mu_{T_{n,i}}, \qquad i - 1, \ldots, N \tag{1.8}$$

where $\mu_{T_{n,i}} = E(T_{n,i})$.
According to equation (1.4) for the credibility factor c_i we have

$$c_{r,i} = \frac{n \cdot \text{Var}(E(T_{n,i})|\vartheta_i)}{n \cdot \text{Var}(E(T_{n,i})|\vartheta_i) + E(\text{Var}(T_{n,i}|\vartheta_i))}, \, i = 1, \ldots, N.$$

Unfortunately, we are again in the situation of the classical credibility model, i.e. we do not know the relevant distributions to calculate $c_{r,i}$. Therefore, we estimate all necessary parameters yielding the *empirical robust credibility premium*

$$\hat{\mu}_{rC,i;\, n} = \hat{c}_{r,i} \cdot T_{n,i} + (1 - \hat{c}_{r,i}) \cdot \hat{\mu}_{T_{n,i}},$$

$$\hat{\mu}_{T_{n,i}} = \hat{E}(T_{n,i})$$

$$\hat{c}_{r,i} = \frac{n \cdot \widehat{\text{Var}}(E(T_{n,i})|\vartheta_i)}{n \cdot \widehat{\text{Var}}(E(T_{n,i})|\vartheta_i) + \hat{E}(\text{Var}(T_{n,i}|\vartheta_i))}.$$

Remark 1.4
In the Bühlmann credibility premium (1.3), the class mean μ_{coll} was equal to

$$\mu_{coll} = E(\mu(\vartheta_1))$$

due to the i.i.d. random variables $\vartheta_1, \ldots, \vartheta_N$ and the conditionally i.i.d. random variables $X_{i1}, \ldots, X_{in}$.
As we will point out in Chapter 5, the robust estimator $T_{n,i}$ is not the sum of independent random variables any more. Thus, our class mean depends on the risk class as well.

Remark 1.5
Note that the actual purpose of a robust estimator is to identify and downgrade outliers that deviate from the true model.
In our case, extreme values are caused by the model itself and we employ robust estimators to identify these extreme values.

Now we turn our attention to μ_{xs}. By determining the estimate value of $T_{n,i}$, we are able to identify those claim amounts that contain extreme parts. These extreme parts are going to be used for estimating μ_{xs}.

There is an instrument from Explorative Data Analysis that measures exceedances of predefined thresholds. It is called the *mean excess function*.

Let ℓ be the threshold above which we speak of excess claim amounts, and denote by X the claim amount, then the mean excess function is

$$e(\ell) = E(X - \ell | X > \ell).$$

Because we are assumimg the same contamination for all risk classes, cf. model (1.6), our estimator will be

$$\hat{e}(\ell) = \frac{1}{\sum\limits_{i=1}^{N}\sum\limits_{j=1}^{n} \mathbb{1}_{(\ell,\infty)}(X_{ij})} \cdot \sum\limits_{i=1}^{N}\sum\limits_{j=1}^{n} \mathbb{1}_{(\ell,\infty)}(X_{ij}) \cdot (X_{ij} - \ell). \tag{1.9}$$

We continue with some notes about our assumptions on the model distribution functions. As Rolski et al. [Rolski et al. 1999], p. 71, points out, the Pareto distribution function with shape parameter $\lambda > 0$ is quite common to be applied when dealing with extreme events. We will follow this idea and use the Pareto distribution with parametrisation

$$f(x) = \frac{\lambda \cdot x_0^{\lambda}}{x^{\lambda+1}}, \qquad x > x_0, \qquad x_0, \lambda > 0.$$

Note that the Pareto distribution function belongs to the class of heavy-tailed distribution functions with the expected value existing for $\lambda > 1$ and the variance existing for $\lambda > 2$, cf. Müller [Müller 1991] and Embrechts et al. [Embrechts et al. 2003]. The value of x_0 used for our purposes will be given at a later stage. To find out more about the extreme value distribution functions or the Pareto distribution function in particular, we refer to Resnick [Resnick 1987] and again Johnson et al. [Johnson et al. 1994].

Moreover, the Gamma distribution $\Gamma(\alpha, \theta)$ is commonly used in non-life-insurance, as mentioned by Rolski et al. [Rolski et al. 1999], because the class of Gamma distributions is closed under summation. It occurs for example when talking about homogeneous Poisson processes. Detailed work on the latter has been done by Mikosch [Mikosch 2004]. Hence we choose $F_{\vartheta_i} = F_{\theta_i}$ to be a Gamma distribution function $\Gamma(\alpha, \theta_i)$ with density

$$f_{\theta_i}(x) = \frac{1}{\Gamma(\alpha)\theta_i^{\alpha}} \cdot x^{\alpha-1} e^{x/\theta_i}, \qquad x > 0$$

with α being the shape and θ_i the scale parameter, $i = 1, \ldots, N$.

For $X \sim \Gamma(\alpha, \theta)$ with the above parametrisation we have $E(X) = \alpha\theta$ and $Var(X) = \alpha\theta^2$. Since θ is the scale parameter, we have

$$f_{\theta}(x) = \frac{1}{\theta} \cdot f_1\left(\frac{x}{\theta}\right).$$

Putting everything together, our ε-contaminated statistical model is

$$\mathcal{F}_{\varepsilon}(\Gamma(\alpha, \theta_i)) = \{F \colon F = (1 - \varepsilon) \cdot \Gamma(\alpha, \theta_i) + \varepsilon \cdot Par(\lambda, x_0)\} \tag{1.10}$$

for given α, x_0, λ and ε. Note that, to keep things simple, we denote both the Gamma distribution and its distribution function by $\Gamma(\alpha, \theta)$.

The scale parameter θ_i of the Gamma distribution will be the risk parameter. Thus, from now on we write θ_i instead of ϑ_i.

We conclude the chapter with a summary of our model. Let $\mathcal{F}$ be the set of continuous distribution functions.

(M1) The risk portfolio consists of N risk classes, each of them associated with a risk parameter $\theta_i \in \Theta$, $i = 1, \ldots, N$. The risk parameters $\theta_1, \ldots, \theta_N$ are i.i.d. random variables with structure function $U \in \mathcal{F}$.

(M2) In each risk class i there are n claim amounts denoted by X_{ij}, $j = 1, \ldots, n$. The claim amounts X_{ij} are conditionally i.i.d. for given θ_i with distribution function F_{θ_i}. The claim amount distribution function F_{θ_i} for risk class i is an element of the ε-contamination class (1.10).

(M3) The individual premium $\mu_{\text{ind},i}$ for each risk class consists of an ordinary part $\mu_{\text{ord},i}$ and an extreme part μ_{xs}, in which the ordinary part depends on the risk parameter θ_i,

$$\mu_{\text{ind},i} = \mu_{\text{ord},i} + \mu_{\text{xs}} = \mu(\theta_i) + \mu_{\text{xs}}.$$

(M4) The premium $\mu_{\text{ord},i}$ is estimated by the empirical robust credibility premium

$$\hat{\mu}_{\text{rC},i} = \hat{c}_{r,i} \cdot T_{n,i} + (1 - \hat{c}_{r,i}) \cdot \mu_{T_{n,i}}, \quad \hat{c}_{r,i} = \frac{n \cdot \widehat{\text{Var}}(E(T_{n,i})|\theta_i)}{n \cdot \widehat{\text{Var}}(E(T_{n,i})|\theta_i) + \hat{E}(\text{Var}(T_{n,i}|\theta_i))}.$$

(M5) The excess premium μ_{xs} is estimated by the empirical mean excess function (1.9)

$$\hat{e}(\ell) = \frac{1}{\sum\limits_{i=1}^{N} \sum\limits_{j=1}^{n} \mathbb{1}_{(\ell,\infty)}(X_{ij})} \cdot \sum_{i=1}^{N} \sum_{j=1}^{n} \mathbb{1}_{(\ell,\infty)}(X_{ij}) \cdot (X_{ij} - \ell).$$

We continue our work keeping (M1)-(M5) in our minds.

Chapter 2

Robust Estimation

As we pointed out in Chapter 1, the focus of this thesis lies on quantitative robustness. This chapter deals with certain aspects of quantitative robustness that are needed in the following chapters. Especially, we will turn our attention to M-estimators and their properties.

From now on X is a random variable on a probability space $(\Omega, \mathfrak{A}, P)$ taking values in $(\Omega', \mathfrak{A}') = (\mathbb{R}, \mathfrak{B})$.

2.1 Estimators as Functionals

Let X be a random variable with distribution P_X on $(\mathbb{R}, \mathfrak{B})$. We assume further that P_X can be parametrised by some parameter $\theta \in \Theta \subseteq \mathbb{R}$. To emphasise the parametrisation, we write $P_{X;\theta}$.
By $\mathbf{Q}_\Theta$ denote the set

$$\mathbf{Q}_\Theta = \big\{ Q_\theta : \theta \in \Theta, \, Q_\theta \text{ is distribution on } (\mathbb{R}, \mathfrak{B}) \big\}$$

of distributions parametrised by $\theta \in \Theta$. Considering the parametric statistical model $(\mathbb{R}, \mathfrak{B}, \mathbf{Q}_\Theta)$, cf. Definition 1.1, we assume, there exists a $\theta_0 \in \Theta$ such that $P_{X;\theta} = Q_{\theta_0}$ with $Q_{\theta_0} \in \mathbf{Q}_\Theta$.

A common way to introduce an estimator δ of $\tau(\theta)$, $\tau : \Theta \to \mathbb{R}$ is to define δ as a function from $\mathbb{R}$ to $\mathbf{Q}_\Theta$, (cf. Müller [Müller 1991]).
But in this work we will use another definition given by both Hampel [Hampel 1971] and Jurečková [Jurečková, Picek 2006]). Note that for each probability distribution $Q_\theta \in \mathbf{Q}_\Theta$ there is a unique distribution function $F : \mathbb{R} \to [0,1]$ defined by

$$F_\theta(x) := Q_\theta((-\infty, x]).$$

It is therefore obvious that a distribution $P_{X;\theta}$ that is parametrised by $\theta \in \Theta$ leads to a distribution function F_θ that is also parametrised by the parameter $\theta \in \Theta$.
Equivalently, for an empirical distribution $Q(\cdot; x_1, \ldots, x_n)$ of a random sample $X_1, \ldots, X_n$,

the empirical distribution function F_n is determined through

$$F_n(x; x_1, \ldots, x_n) := Q_n((-\infty, x]; x_1, \ldots, x_n), \quad x \in \mathbb{R}$$

and we will write $F_{n;\theta}$ when it is necessary to show the dependence on θ; i.e. $F_{n;\theta}$ is the empirical distribution function of some random sample of size n from F_θ.
From now on, we assume $\tau(\theta) = \theta$ and the set of distribution functions belonging to our parametric statistical model $(\mathbb{R}, \mathfrak{B}, \mathbf{Q}_\Theta)$ is denoted by $\mathcal{F}_\Theta$.

Definition 2.1
As above, X is a random variable with distribution function F_θ. Denote by $\mathcal{F}_n$ the set of all empirical distribution functions of size n.
Let T be a functional that maps the set of empirical distributions functions into the real numbers,

$$T : \mathcal{F}_n \to \mathbb{R}, \qquad F_n \mapsto T(F_n),$$

then $T(F_n)$ is called an estimator *of θ.*

We will denote $T(F_n)$ briefly by T_n, if it is unambiguous which empirical distribution function is meant.
Note that – as pointed out by Hampel [Hampel 1971] for example – by using the empirical distribution function instead of the data itself, we may ignore some important information such as the interdependence within the random sample.

Let T be a functional on $\mathcal{F}$, the set of all distribution functions. Then the parameter $T(F_\theta)$ determined by T can be estimated using $T(F_{n;\theta})$. Note that $T(F_\theta)$ not necessarily has to be equal to θ but surely is some function of θ.

If $T(F_{n;\theta}) \xrightarrow{P} T(F_\theta)$ then T_n is a *consistent estimator*. If in addition $T(F_\theta) = \theta$, then T_n is a *Fisher-consistent* estimator. We will approach both the convergence and the condition $T(F_\theta) = \theta$ at a later stage.
Here we only mention, that T_n being a consistent estimator in fact means, the functional T is continuous at F with respect to the weak*-topology, cf. Huber [Huber 1981].

At this point we go back to qualitative robustness for a moment. If T is a functional with $T_n \xrightarrow{P} T(F)$, then the robustness of the sequence of estimators $(T_n)_{n\in\mathbb{N}}$ as given in Section 1.3 follows from the weak*-continuity of the functional T applying the Prohorov distance, cf. Hampel [Hampel 1971], Theorem 1a and the Corollary.
We now turn to a certain type of estimators that is often used when dealing with robust estimation.

2.2   *M*-Estimators

Again we consider the situation from the beginning of Section 2.1. Our aim is to estimate the parameter $\theta \in \Theta$ of some distribution function $F_\theta \in \mathcal{F}_\theta$. A well known classical estimator would be the *maximum likelihood estimator*. It is defined as

$$\theta_{\mathrm{ML}} = \underset{\theta \in \Theta}{\operatorname{argmin}}\, L(\theta; X_1, \ldots, X_n),$$

where $L(\theta; x_1, \ldots, x_n)$ is the likelihood function, cf. Müller [Müller 1991]

$$L(\theta; x_1, \ldots, x_n) = \prod_{i=1}^{n} f_\theta(x_i), \qquad f_\theta(x) = \frac{d}{dx} F_\theta(x)$$

with respect to the random sample $X_1, \ldots, X_n$ drawn from F_θ.

If F_θ is continuous with density f_θ, we often work with the *log*−likelihood function

$$l(\theta; X_1, \ldots, X_n) = \ln(L(\theta; X_1, \ldots, X_n)) = \sum_{i=1}^{n} \ln f_\theta(X_i).$$

It is obvious that

$$\theta_{\mathrm{ML}} = \underset{\theta \in \Theta}{\operatorname{argmin}}\, L(\theta; X_1, \ldots, X_n) \iff \theta_{\mathrm{ML}} = \underset{\theta \in \Theta}{\operatorname{argmin}}\, l(\theta; X_1, \ldots, X_n).$$

Especially in the context of point estimation in linear models, Gauss' least-squared estimation is of some importance, cf. Lehmann, Casella [Lehmann, Casella 1998]. There, the *minimum least square estimator* θ_{MLS} is defined as

$$\theta_{\mathrm{MLS}} = \underset{\theta \in \Theta}{\operatorname{argmin}} \sum_{i=1}^{n} (X_i - \theta)^2. \tag{2.1}$$

Even though the least-squares approach is easy to handle, it is very sensitive towards outliers.

Therefore, Huber suggested in [Huber 1964] to look for estimators that are defined similarly to (2.1) but are more robust towards outliers.

Definition 2.2
Let $\rho : \mathbb{R} \times \Theta \to \mathbb{R}$ be a measurable function and $X_1, \ldots, X_n$ be a random sample from distribution function F_θ.
The M-estimator θ_M is the solution of

$$\theta_M = \underset{\theta \subset \Theta}{\operatorname{argmin}} \sum_{i=1}^{n} \rho(X_i; \theta). \tag{2.2}$$

In cases where the partial derivative of ρ with respect to θ exists, the solution θ_M of (2.2) also fulfils

$$\sum_{i=1}^{n} \psi(X_i; \theta_M) = 0, \qquad \psi(x; \theta) := \frac{\partial}{\partial \theta} \rho(x; \theta). \tag{2.3}$$

From now on, we use the definition of an *M*-estimator as given in equation (2.3).

Remark 2.3

The function ρ in equation (2.2) has to be strictly convex in θ to ensure that every solution of (2.3) is as well a solution of (2.2). In this case the solution is unique, cf. Rockafellar [Rockafellar 1972].

Example

Let F_θ be a continuous distribution function with density f_θ and choose

$$\rho(x; \theta) = \ln f_\theta(x).$$

Then

$$\psi(x; \theta) = \frac{\frac{\partial}{\partial \theta} f_\theta(x)}{f_\theta(x)}$$

and θ_{ML} solving the following equation for θ

$$\sum_{i=1}^{n} \psi(X_i; \theta) = 0$$

is the common maximum likelihood estimator.

Resuming the concept of Section 2.1, let $T : \mathcal{F}_n \to \mathbb{R}$ be a functional such that θ_M in (2.3) can be written $\theta_M = T_n$, then (2.3) means

$$\int_{\mathbb{R}} \psi(s; T_n) dF_n(s) = 0. \tag{2.4}$$

Note that Huber in [Huber 1964] did not define robust estimators as functionals of the empirical distribution function. He applied the classical definition of estimators being functions on the sample space.

The interpretation of the estimator as a functional on the set of empirical distribution functions is for example used by Hampel [Hampel 1971], [Hampel 1974] and later on by Künsch [Künsch 1992] and Gisler, Reinhard ([Gisler, Reinhard 1993]).

2.2.1   Definition of a Location M-Estimator

Let F_θ be a distribution function with parameter $\theta \in \Theta \subseteq \mathbb{R}$. We call θ a *location parameter* if $F_\theta(x) = F_0(x - \theta)$ for all $\theta \in \Theta$.

A statistic $\delta_n = \delta(X_1, \dots, X_n)$ is called *location equivariant* if

$$\delta(X_1 + a, \dots, X_n + a) = \delta(X_1, \dots, X_n) + a,$$

cf. Lehmann, Casella [Lehmann, Casella 1998].

We transfer this concept into our context by saying a functional T is *location equivariant* if

$$T(F_\theta) = T(F_0) + \theta, \qquad \text{for all } \theta \in \Theta.$$

It should be obvious, that a reasonable demand on an estimator T_n of a location parameter θ is its location equivariance.

So let T_n be an M-estimator according to equation (2.4). As it has already been suggested by Huber in [Huber 1964] and later mentioned in numerous other papers such as Gisler, Reinhard [Gisler, Reinhard 1993], in the case of estimating θ being a location parameter, the function $\psi : \mathbb{R} \times \Theta \to \mathbb{R}$ in (2.4) is replaced by a function $\hat{\psi} : \mathbb{R} \to \mathbb{R}$ with $\hat{\psi}(x - \theta) := \psi(x; \theta)$.

Lemma 2.4
Let F_θ be a distribution function with location parameter $\theta \in \Theta$. If the solution $T(F_\theta)$ of

$$\int_{\mathbb{R}} \hat{\psi}(s - T(F_\theta)) dF_\theta(s) = 0 \tag{2.5}$$

is unique, the functional T is location equivariant.

Proof: If X is a random variable with distribution function F_θ, θ being a location parameter, then the distribution function G of the random variable $Y = X + a, a \in \mathbb{R}$ has location parameter $\theta + a$, since $G(y) = F_\theta(y - a) = F_0(y - (\theta + a))$. Thus, proving the lemma, it is enough to show $T(F_\theta) = T(F_0) + \theta$ for all $\theta \in \Theta$.

$$0 = \int_{\mathbb{R}} \hat{\psi}(s - T(F_\theta)) dF_\theta(s) \overset{z=s-\theta}{=} \int_{\mathbb{R}} \hat{\psi}(z + \theta - T(F_\theta)) dF_\theta(z + \theta)$$

$$= \int_{\mathbb{R}} \hat{\psi}(z - (T(F_\theta) - \theta)) dF_0(z).$$

In addition we know

$$\int_{\mathbb{R}} \hat{\psi}(z - T(F_0)) dF_0(z) = 0.$$

Thus, from the uniqueness of the solution follows $T(F_\theta) = T(F_0) + \theta$ for all $\theta \in \Theta$. □

Indeed, Lemma 2.4 means that any location equivariant functional T is sufficiently described by

$$\int_{\mathbb{R}} \hat{\psi}(s - T(F_0)) dF_0(s) = 0. \tag{2.6}$$

As it has already been mentioned, Huber demonstrated his ideas of robust statistics at a location M-estimator ([Huber 1964]) without employing a functional T on the set of distribution functions.

2.2.2 Definition of a Scale M-Estimator

A parameter $\theta \in \Theta = (0, \infty)$ of a distribution function F_θ is called *scale parameter*, if $F_\theta(x) = F_1(x/\theta)$ for all $\theta \in \Theta$.
Again, a statistic $\delta_n = \delta(X_1, \ldots, X_n)$ is called *scale equivariant* if

$$\delta(aX_1, \ldots, aX_n) = a \cdot \delta(X_1, \ldots, X_n), \qquad a > 0$$

and we will call T *scale equvariant* if

$$T(F_\theta) = \theta \cdot T(F_1), \qquad \text{for all } \theta \in \Theta.$$

As in the case of estimating a location parameter, it is reasonable to ask for a scale equivariant functional when estimating a scale parameter θ.
Similar to the situation in Section 2.2.1 we substitute a function $\hat{\psi} : (0, \infty) \to \mathbb{R}$ for the function ψ in (2.4) such that

$$\hat{\psi}\left(\tfrac{x}{\theta}\right) := \psi(x; \theta), \tag{2.7}$$

cf. Thall [Thall 1979], Künsch [Künsch 1992].
Before proving that equation (2.7) leads to a scale equivariant functional, we show the uniqueness of the solution. Unlike Section 2.2.1 we now put restriction on $\hat{\psi}$ rather than ρ.

Lemma 2.5
Let F be a continuous distribution function. If the function $\hat{\psi} : (0, \infty) \to \mathbb{R}$ is monotonically increasing, then the solution $T(F)$ of

$$\int\limits_0^\infty \hat{\psi}\left(\frac{s}{T(F)}\right) dF(s) = 0,$$

is unique a.s.

Proof: Let $T_1, T_2 : \mathcal{F} \to \mathbb{R}$ be such that

$$\int\limits_0^\infty \hat{\psi}\left(\frac{s}{T_1(F)}\right) dF(s) = 0, \qquad \int\limits_0^\infty \hat{\psi}\left(\frac{s}{T_2(F)}\right) dF(s) = 0.$$

We will show that then $T_1(F) = T_2(F)$ almost surely.

Define the sets $\mathcal{S}_= \subset (0, \infty)$ and $\mathcal{S}_{\neq} \subseteq (0, \infty)$ to be

$$\mathcal{S}_= = \left\{ s \in (0, \infty) : \hat{\psi}\left(\frac{s}{T_1(F)}\right) = \hat{\psi}\left(\frac{s}{T_2(F)}\right) \right\}$$

$$\mathcal{S}_{\neq} = \left\{ s \in (0, \infty) : \hat{\psi}\left(\frac{s}{T_1(F)}\right) \neq \hat{\psi}\left(\frac{s}{T_2(F)}\right) \right\}.$$

We then can conclude

$$
\begin{aligned}
0 &= \int_0^\infty \hat{\psi}\left(\frac{s}{T_1(F)}\right) dF(s) - \int_0^\infty \hat{\psi}\left(\frac{s}{T_2(F)}\right) dF(s) \\
&= \int_{\mathcal{S}_=} \left(\hat{\psi}\left(\frac{s}{T_1(F)}\right) - \hat{\psi}\left(\frac{s}{T_2(F)}\right)\right) dF(s) + \int_{\mathcal{S}_{\neq}} \left(\hat{\psi}\left(\frac{s}{T_1(F)}\right) - \hat{\psi}\left(\frac{s}{T_2(F)}\right)\right) dF(s) \\
&= \int_{\mathcal{S}_{\neq}} \left(\hat{\psi}\left(\frac{s}{T_1(F)}\right) - \hat{\psi}\left(\frac{s}{T_2(F)}\right)\right) dF(s).
\end{aligned}
$$

Because $\hat{\psi}$ is monotone increasing the argument of the last integral is either positive or negative. But then $\mathcal{S}_{\neq}$ has to be a null set since F is a continuous distribution function.
□

Now we can prove the scale equivariance of the functional T defined by (2.7) and Lemma 2.5.

Lemma 2.6
Let F_θ be a continuous distribution function with scale parameter $\theta \in \Theta$ and let $\hat{\psi}$: $(0,\infty) \to \mathbb{R}$ be monotonically increasing. Then the functional T defined by

$$
\int_{\mathbb{R}} \hat{\psi}\left(\frac{s}{T(F_\theta)}\right) dF_\theta(s) = 0 \tag{2.8}
$$

is scale equivariant.

Proof: Again it is enough to verify $T(F_\theta) = \theta \cdot T(F_1)$ for all $\theta \in \Theta = (0,\infty)$, since for random variables X, Y with $Y = cX, c > 0$ we get

$$
G_Y(y) = F_{X;\theta}\left(\frac{y}{c}\right) = F_{X;1}\left(\frac{y}{c \cdot \theta}\right) = F_{X;1}\left(\frac{y}{\tilde{\theta}}\right), \quad \tilde{\theta} \in \Theta.
$$

Hence for F_θ

$$
0 = \int_{\mathbb{R}} \hat{\psi}\left(\frac{y}{T(F_\theta)}\right) dF_\theta(y) \stackrel{y=\theta s}{=} \int_{\mathbb{R}} \hat{\psi}\left(\frac{\theta s}{T(F_\theta)}\right) dF_\theta(\theta s) = \frac{1}{\theta} \int_{\mathbb{R}} \hat{\psi}\left(\frac{s}{\frac{T(F_\theta)}{\theta}}\right) dF_1(s).
$$

Besides, for F_1 we know

$$
0 = \int_{\mathbb{R}} \hat{\psi}\left(\frac{s}{T(F_1)}\right) dF_1(s)
$$

and because of the uniqueness of the solution this means

$$
T(F_\theta) = \theta \cdot T(F_1). \tag{2.9}
$$
□

It follows from Lemma 2.6 that the scale equivariant functional T is sufficiently described by

$$\int_{\mathbb{R}} \hat{\psi}\left(\frac{s}{T(F_1)}\right) dF_1(s) = 0 \tag{2.10}$$

because of (2.9).

We will give some additional information on the research that has been done on robust scale estimators. Already in [Huber 1964] M-estimators for scale parameters can be found. Thall [Thall 1979] develops a scale estimator for the Exponential distribution by solving a statistical game. Kimber [Kimber 1983] compares different robust estimators for the scale parameter of the Gamma distribution whereas Gather, Schultze [Gather, Schultze 1999] introduce an M-estimator for the scale parameter of the Exponential distribution and compare it to different other estimators having the same breakdown point. Künsch [Künsch 1992] and Gisler, Reinhard [Gisler, Reinhard 1993] give applications of scale M-estimators in insurance mathematics. Collins [Collins 1999] deals with robust M-estimators of scale comparing minimax bias and maximal variance. Szatmari, Collins [Szatmari, Collins 2007] analyse scale M-estimators with minimum gross error sensitivity (2.16) for symmetric and unimodal distributions.

2.2.3   Fisher-consistent M-Estimators

Statistics knows several criteria to judge an estimator, *unbiasedness* being one of them. Because we are dealing with functionals, the concept of unbiasedness is not appropriate as explained by Hampel et al. [Hampel et al. 1986], p. 83. But there is a more general idea, which is similar to unbiasedness, cf. Hampel [Hampel 1974].

Definition 2.7
Let $F_\theta \in \mathcal{F}$ be a distribution function parametrised by $\theta \in \Theta \subseteq \mathbb{R}$ and $T : \mathcal{F} \to \mathbb{R}$ be the functional that determines the estimator T_n of θ. The functional T is called Fisher-consistent, *if*

$$T(F_\theta) = \theta \qquad \text{for all } \theta \in \Theta.$$

Assume, ψ in (2.4) is chosen such that for all $\theta \in \Theta$

$$\int_{\mathbb{R}} \psi(s; \theta) dF_\theta(s) = 0.$$

Then obviously the functional T determined as solution of

$$\int_{\mathbb{R}} \psi(s; T(F_\theta)) dF_\theta(s) = 0 \tag{2.11}$$

is Fisher-consistent if it is unique.

2.3   Main Concepts in Quantitative Robustness

The last section presented a certain class of estimators that was introduced by Huber [Huber 1964] to handle the problem of outlier-sensitivity of popular estimators such as the Gauss' minimum-least-square estimator.
Since qualitative robustness (Section 1.3.1) is quite unpleasant to deal with, we still lack some useful criteria to value and rank the robustness of different estimators.
In Section 1.3.2 we already gave a short overview of important terms in quantitative robustness. We will now have a closer look at these quantities.

A good description of quantitative robustness gives the question: How much impact does a single observation has on the behaviour of the estimator?

Thus, many characteristics in quantitative robustness are defined for finite samples. Donoho, Huber [Donoho, Huber 1983] use the term *corrupted sample* to describe a random sample containing "abnormal" observations. They distinguish three kinds of corruptions

- ε-**contamination:** m arbitrary observations are added to the sample of size n, i.e. the contamination rate is $\varepsilon = \frac{m}{m+n}$

- ε-**replacement:** $m < n$ arbitrary observations are replaced by m (abnormal) observations, i.e. the replacement rate is $\varepsilon = \frac{m}{n}$

- ε-**modification:** Instead of the original random sample with empirical distribution P_n, another sample with empirical distribution Q_m is chosen. It is assumed that Q_m lies in the neighbourhood of P_n.

We will denote the empirical distribution function of the original sample with F_n and the one of the corrupted sample with $\hat{F}_m$. Note that both underlying samples may be of different lengths.

A characteristic that is often easy to calculate, is the *finite sample maximum bias* defined by

$$b(\varepsilon, F_n, T) = \sup_{\hat{F}_m \in \mathcal{F}_{m;\varepsilon}} \left| T(F_n) - T(\hat{F}_m) \right|, \qquad (2.12)$$

where $\mathcal{F}_{m;\varepsilon}$ is the set of empirical distribution functions of the corrupted samples with corruption rate ε.

Based on the finite sample maximum bias Donoho, Huber calculate the *finite sample breakdown point*

$$\varepsilon_n^\star(F_n, T) = \inf_{0 < \varepsilon < 1} \left\{ b(\varepsilon, F_n, T) = \infty \right\}.$$

In general parlance, the finite sample breakdown point specifies the maximum fraction of the data that may be differ widely from the remaining sample without upsetting the estimator.

Now assume that T is a Fisher-consistent functional applied to determine a location M-estimator. In such cases Hampel et al. [Hampel et al. 1986] suggest – in the case of ε-replacement – to calculate the finite sample breakdown point through

$$\varepsilon_n^\star(F_n, T) = \frac{1}{n} \max \left\{ m : T(\hat{F}_m) < \infty, \ \hat{F}_m \in \mathcal{F}_{m,\varepsilon} \right\}.$$

Rousseeuw, Leroy [Rousseeuw, Leroy 1987] adjust this definition for Fisher-consistent scale M-estimators

$$\varepsilon_n^\star(F_n, T) = \frac{1}{n} \min \left\{ m : T(\hat{F}_m) = \infty \text{ or } T(\hat{F}_m) = 0, \ \hat{F}_m \in \mathcal{F}_{m,\varepsilon} \right\}. \tag{2.13}$$

This definition is also used by Gather, Schultze [Gather, Schultze 1999].

It is apparent that $0 < \varepsilon_n^\star \leq 0.5$.
As an example consider the mean. It has a finite sample breakdown point of $\frac{1}{n}$ converging to 0 for infinite sample size n. In contrast, the median has a finite sample breakdown point of $\lfloor \frac{n}{2} \rfloor$ converging to 0.5. Even though half of the data is contaminated, the median gives a reasonable estimate.

Of course the concepts of finite sample maximum bias and finite sample breakdown point can be generalised. To do so, we need a generalisation of an ε-corrupted sample first. The ε-contamination model (1.6) in Section 1.4 is a natural generalisation of both the ε-contamination and ε-replacement introduced above. Keeping this in mind we define the *maximum bias*

$$b(\varepsilon, F, T) = \sup_{\hat{F} \in \mathcal{F}_\varepsilon} \left| T(F) - T(\hat{F}) \right|,$$

cf. Huber [Huber 1981]. Unfortunately, the generalisation of the finite sample breakdown point is not as easy.

The concept of the breakdown point introduced by Hampel [Hampel 1971] is settled in qualitative robustness and uses the Prohorov distance of Definiton 1.2. Hampel defines the breakdown point $\varepsilon^\star(P, (T_n)_{n \in \mathbb{N}})$ of a sequence of estimators $(T_n)_{n \in \mathbb{N}}$ with respect to the true distribution P by

$$\varepsilon^\star(P, (T_n)_{n \in \mathbb{N}}) = \sup\{\varepsilon \in (0, 1] : \exists \text{ compact } K_\varepsilon \subseteq \Theta :$$
$$d_{\mathrm{Proh}}(P, Q) < \varepsilon \Rightarrow \lim_{n \to \infty} Q(\{T_n \in K_\varepsilon\}) = 1\}.$$

Another important concept in quantitative robustness is the *influence function*. We recall that ε-contamination in a finite sample means adding m additional observations to the sample. Now, the influence function tells us how much impact one single observation has on the estimator. Again, we start with the finite sample version that is also called *sensitivity curve*, cf. Tukey [Tukey 1970] cited in Andrews et al. [Andrews et al. 1972], Hogg [Hogg 1979] and Gisler, Reinhard [Gisler, Reinhard 1993].

$$SC(x; F_n, T) = \frac{T\left(\frac{n-1}{n} F_{n-1} + \frac{1}{n}\Delta_x\right)}{\frac{1}{n}},$$

where Δ_x is the degenerate distribution function with point mass 1 at x.

The sensitivity curve can be seen as a discretised version of the influence function. In order to define the influence function and for later purposes we need the following two definitions.

Definition 2.8
Let X, Y be Banach spaces, $U \subset X$ an open set and let T be an operator, $T : U \to Y$. Consider $x \in U$. If for $h \in Y$, $t \in \mathbb{R}$ the following limit exists

$$\lim_{t \to 0} \frac{T(x + th) - T(x)}{t} =: T'_h(x),$$

then T is called differentiable at x in direction h and $T'_h(x)$ is the derivative.
If in addition $T'_h(x)$ is continuous and linear in h, then it is called Gâteaux-derivative.

Note that in the above definition, $t \in \mathbb{R}$ has to be chosen such that $x + th \in U$.

Now, if $y \in U$, put $h := y - x$. Then Definition 2.8 means that the directional derivative of the operator T is the ordinary derivative of the real function $\bar{H}(t) := T((1 - t)x + ty)$ at $t = 0$.

Next, we consider distribution function F, $G \in \mathcal{F}$ such that $F + t(G - F) = (1 - t)F + tG$ lies in some neigbourhood $U(F)$ for $|t| \leq \varepsilon(G)$. Then the directional derivative of T at distribution function F in direction $G - F$ is

$$\lim_{t \to 0} \frac{T((1 - t)F + tG) - T(F)}{t} = \frac{d}{dt} T((1 - t)F + tG)\Big|_{t=0} = T'_{G-F}(F), \qquad (2.14)$$

if the limit exists.

Choosing $G = \Delta_x$, we define the influence function of T at F to be

$$IF(x; F, T) :- \lim_{t \to 0+} \frac{T((1 - t)F + t\Delta_x) - T(F)}{t}. \qquad (2.15)$$

Note that the influence function is not the Gâteaux-derivative in the sense of definition 2.8, because $\mathcal{F}$ is not a Banach space.
However, for reasons of brevity, we will call $IF(x; F, T)$ the Gâteaux-derivative keeping the discrepancies in mind.

Definition 2.9
Let X, Y be Banach spaces and $U \subset X$ an open set. An operator $T : U \to Y$ is called totally differentiable in $x \in U$ if there exists a bounded linear mapping $A[x] : U \to Y$ such that

$$T(x + h) = T(x) + A[x](h) + o(h), \qquad \lim_{h \to 0} o(h) = 0.$$

Again disobeying the fact that $\mathcal{F}$ is not a Banach space, we call T totally differentiable (or *Fréchet-differentiable*) in $F \in \mathcal{F}$ if for $G \in \mathcal{F}$ there exists a bounded linear map $A[F]$ on the space of finite signed measures, see Huber [Huber 1981], p.35, such that

$$T(F + G) = T(F) + A[F](G - F) + o(|G - F|).$$

Sometimes the Fréchet derivative is also called *strong derivative*. It is well known from analysis, that any Fréchet-differentiable function is Gâteaux-differentiable, whereas the converse is not true.

If T is Gâteaux-differentiable, then by (2.14)

$$IF(x; F, T) = T'_{\Delta_x - F}(F),$$

cf. Hampel [Hampel 1974].

The maximum influence an additional observation can have, is given by the *gross error sensitivity*

$$\gamma^*(T) = \sup_{x \in \mathbb{R}} |IF(x; F, T)|, \tag{2.16}$$

introduced by Hampel [Hampel 1974]. Hampel also points out that a robust estimator should have a bounded gross error sensitivity.

At this point we refer to the book of Hampel et al. [Hampel et al. 1986]. The authors give a detailed overview of robust statistics, its history and different concepts, and explain the difficulties of finding estimators that are robust in more than one sense.

In a next step, we will calculate the influence function of M-estimators in general and use it to define a certain robust M-estimator.

2.3.1 Influence Functions of M-Estimators

Recall from Section 2.2 the definition of an M-estimator through (2.2)

$$\theta_M = \operatorname*{argmin}_{\theta \in \Theta} \sum_{i=1}^{n} \rho(X_i; \theta).$$

For the remaining part of this chapter we will assume ρ to be strictly convex, such that the M-estimator $\theta_M = T_n$ is also uniquely definied by (2.3)

$$\sum_{i=1}^{n} \psi(X_i; T_n) = 0, \qquad \psi(x; \theta) = \frac{\partial}{\partial \theta} \rho(x; \theta), \qquad \text{for all } \theta \in \Theta.$$

We also assume $T_n \xrightarrow{P} T(F_\theta)$.

If the function $\psi : \mathbb{R} \times \Theta \to \mathbb{R}$ in (2.3) is partially differentiable with respect to $\theta \in \Theta$ and T is Gâteaux-differentiable, the influence function (2.15) can be expressed in terms of ψ. To see this, let $\hat{H} : \mathbb{R} \times \mathcal{F}^{\mathbb{R}} \to \mathbb{R}$ be the function

$$\hat{H}(t, T) = \int_{\mathbb{R}} \psi(s; T((1-t)F_\theta + t\Delta_x)) d((1-t)F_\theta + t\Delta_x)(s) \tag{2.17}$$

and T be the functional solving

$$\hat{H}(0, T) = \int_{\mathbb{R}} \psi(s; T(F_\theta)) dF_\theta(s) = 0. \tag{2.18}$$

Now we differentiate (2.17) on both sides with respect to t and use the interchange of differentiation and integration, cf. Mangoldt, Knopp [Mangoldt, Knopp 1975] to get

$$\frac{\partial}{\partial t} \hat{H}(t, T) = \frac{\partial}{\partial t} \left[(1-t) \int_{\mathbb{R}} \psi(s; T((1-t)F_\theta + t\Delta_x)) dF_\theta(s) \right]$$

$$+ \frac{\partial}{\partial t} \left[t \int_{\mathbb{R}} \psi(s; T((1-t)F_\theta + t\Delta_x)) d\Delta_x(s) \right]$$

$$= - \int_{\mathbb{R}} \psi(s; T((1-t)F_\theta + t\Delta_x)) dF_\theta(s)$$

$$+ (1-t) \int_{\mathbb{R}} \frac{\partial}{\partial \vartheta} \psi(s; \vartheta) \Big|_{\vartheta = T((1-t)F_\theta + t\Delta_x)} \cdot \frac{d}{dt} T((1-t)F_\theta + t\Delta_x) dF_\theta(s)$$

$$+ \int_{\mathbb{R}} \psi(s; T((1-t)F_\theta + t\Delta_x)) d\Delta_x(s)$$

$$+ t \int_{\mathbb{R}} \frac{\partial}{\partial \vartheta} \psi(s; \vartheta) \Big|_{\vartheta = T((1-t)F_\theta + t\Delta_x)} \frac{d}{dt} T((1-t)F_\theta + t\Delta_x) d\Delta_x(s)$$

$$- \int_{\mathbb{R}} \psi(s, T((1-t)F_\theta + t\Delta_x)) d(\Delta_x - F_\theta)(s)$$

$$+ (1-t) \int_{\mathbb{R}} \frac{\partial}{\partial \vartheta} \psi(s; \vartheta) \Big|_{\vartheta = T((1-t)F_\theta + t\Delta_x)} \cdot \frac{d}{dt} T((1-t)F_\theta + t\Delta_x) dF_\theta(s)$$

$$+ t \int_{\mathbb{R}} \frac{\partial}{\partial \vartheta} \psi(s; \vartheta) \Big|_{\vartheta = T((1-t)F_\theta + t\Delta_x)} \cdot \frac{d}{dt} T((1-t)F_\theta + t\Delta_x) d\Delta_x(s).$$

Evaluating the above expression at $t = 0$, the first part of the sum equals

$$\int_{\mathbb{R}} \psi(s; T((1-t)F_\theta + t\Delta_x))d(\Delta_x - F_\theta)(s) = \psi(x; T(F_\theta))$$

due to (2.18). Furthermore from equations (2.14) and (2.15) we know

$$\frac{d}{dt}T((1-t)F_\theta + t\Delta_x)\big|_{t=0} = IF(x; F_\theta, T).$$

Thus, the second summand of $(\partial/\partial t)\hat{H}(t, T)$ at $t = 0$ is

$$\int_{\mathbb{R}} \frac{\partial}{\partial\vartheta}\psi(s; \vartheta)\big|_{\vartheta=T(F_\theta)} \cdot \frac{\partial}{\partial t}T((1-t)F_\theta + t\Delta_x)\big|_{t=0}dF_\theta(s)$$

$$= IF(x; F_\theta, T) \cdot \int_{\mathbb{R}} \frac{\partial}{\partial\vartheta}\psi(s; \vartheta)\big|_{\vartheta=T(F_\theta)}dF_\theta(s).$$

Putting everything together, we get

$$\frac{\partial}{\partial t}\hat{H}(t, T)\Big|_{t=0} = \psi(x; T(F_\theta)) + IF(x; F_\theta, T) \cdot \int_{\mathbb{R}} \frac{\partial}{\partial\vartheta}\psi(s; \vartheta)\big|_{\vartheta=T(F_\theta)}dF_\theta(s) = 0,$$

that is

$$IF(x; F_\theta, T) = -\frac{\psi(x; T(F_\theta))}{\int_{\mathbb{R}} \frac{\partial}{\partial\vartheta}\psi(s; \vartheta)\big|_{\vartheta=T(F_\theta)}dF_\theta(s)}. \tag{2.19}$$

it follows that for M-estimators, the influence function is directly proportional to ψ.

Especially for location and scale M-estimators we are able to simplify the influence functions even further, since we know the form of the functions ψ_{loc} and ψ_{scale}. As explained in Section 2.2.1, for a location estimator $\theta_{M;\text{loc}} = T_{\text{loc}}(F_{n;\theta})$ the defining function ψ_{loc} is given by $\psi_{\text{loc}} : \mathbb{R} \to \mathbb{R}$, $\psi_{\text{loc}}(x - \theta) := \psi(x; \theta)$. Hence the influence function is

$$IF(x; F_\theta, T_{\text{loc}}) = \frac{\psi_{\text{loc}}(x - T_{\text{loc}}(F_\theta))}{\int_{\mathbb{R}} \psi'_{\text{loc}}(s - T_{\text{loc}}(F_\theta))dF_\theta(s)}.$$

In the same way, cf. Section 2.2.2, for a scale M-estimator $\theta_{M;\text{scale}} = T_{\text{scale}}(F_\theta)$ the function ψ_{scale} is given by $\psi_{\text{scale}} : \mathbb{R} \to \mathbb{R}$, $\psi_{\text{scale}}\left(\frac{x}{\theta}\right) := \psi(x; \theta)$ and thus

$$IF(x; F_\theta, T_{\text{scale}}) = \frac{\psi_{\text{scale}}\left(\frac{x}{T_{\text{scale}}(F_\theta)}\right)T_{\text{scale}}(F_\theta)}{\int_{\mathbb{R}} \psi'_{\text{scale}}\left(\frac{s}{T_{\text{scale}}(F_\theta)}\right) \cdot \frac{s}{T_{\text{scale}}(F_\theta)}dF_\theta(s)} \tag{2.20}$$

We now state some properties of the influence funcions.

2.3.2   Some Properties of Influence Functions

We just learned that the influence functions of location and scale M-estimators are of special forms. Besides we know from (2.6) and (2.10) that these estimators are defined by

$$\int_{\mathbb{R}} \psi_{loc}(s - T_{loc}(F_0))dF_0(s) = 0 \quad \text{and} \quad \int_0^\infty \psi_{scale}\left(\frac{s}{T_{scale}(F_1)}\right)dF_1(s) = 0.$$

Now we show that the influence functions $IF(x; F_\theta, T_{loc})$ and $IF(x; F_\theta, T_{scale})$ can be described by $IF(x; F_0, T_{loc})$ and $IF(x; F_1, T_{loc})$ provided T_{loc} and T_{scale} are Fisher-consistent,

$$IF(x; F_\theta, T_{loc}) = \frac{\psi_{loc}(x - T_{loc}(F_\theta))}{\int_{\mathbb{R}} \psi'_{loc}(s - T_{loc}(F_\theta))dF_\theta(s)}$$

$$= \frac{\psi_{loc}(x - \theta)}{\int_{\mathbb{R}} \psi'_{loc}(s - \theta)dF_\theta(s)} \qquad \text{due to } T(F_\theta) = \theta$$

$$= \frac{\psi_{loc}(y)}{\int_{\mathbb{R}} \psi'_{loc}(t)dF_0(t)} = IF(y; F_0, T_{loc})$$

where $y = x - \theta$. And

$$IF(x; F_\theta, T_{scale}) = \frac{\psi_{scale}\left(\frac{x}{T_{scale}(F_\theta)}\right)T_{scale}(F_\theta)}{\int_{\mathbb{R}} \psi'_{scale}\left(\frac{s}{T_{scale}(F_\theta)}\right)\frac{s}{T_{scale}(F_\theta)}dF_\theta(s)}$$

$$= \frac{\psi_{scale}\left(\frac{x}{\theta}\right)\theta}{\int_{\mathbb{R}} \psi'_{scale}\left(\frac{s}{\theta}\right)\frac{s}{\theta}dF_\theta(s)} \qquad \text{due to } T_{scale}(F_\theta) = \theta$$

$$= \frac{\psi(y)\theta}{\int_{\mathbb{R}} \psi'_{scale}(t)tdF_1(t)} = \theta \cdot IF(y; F_1, T(F_1)) \qquad (2.21)$$

with $y = x/\theta$. Besides for M-estimators in general

$$E(IF(X; F_\theta, T)) = -\frac{\int_{\mathbb{R}} \psi(s; T(F_\theta))dF_\theta(s)}{\frac{\partial}{\partial \vartheta}\int_{\mathbb{R}} \psi(s; \vartheta)dF_\theta(s)\big|_{\vartheta = T(F_\theta)}} = 0, \qquad (2.22)$$

$$\text{Var}(IF(X; F, T)) = E(IF^2(X; F, T)) \qquad (2.23)$$

if the denominator of (2.22) is non-zero.

With the above results for the influence functions of location and scale M-estimators we immediately get

$$\text{Var}(IF(X; F_\theta, T_{loc})) = E(IF^2(X; F_\theta, T_{loc})) = E(IF^2(X - \theta; F_0, T_{loc}))$$

$$\text{Var}(IF(X; F_\theta, T_{scale})) = \text{Var}(\theta \cdot IF(X/\theta; F_1, T_{scale})) = \theta^2 \cdot E(IF^2(X/\theta; F_1, T_{scale})).$$

$$(2.24)$$

2.3.3   The Asymptotic Variance of T_n

By now we learned a few things about M-estimators and their quantitative characteristics.

With regard to the considerations of the next section, where we will look for an optimal robust M-estimator, we now turn our attention to the asymptotic variance of an M-estimator T_n of a parameter θ.
We restrict ourselves to the case of θ being a scale parameter and because of Lemma 2.6 and (2.10) we only consider the case $\theta = 1$. Furthermore we assume the operator T to be linear, cf. Rudin [Rudin 1991].

Let T be Gâteaux-differentiable in F_1, cf. Definition 2.8 and consider the distribution function

$$H_t = (1 - t)F_1 + t\Delta_x, \qquad x \in \mathbb{R}.$$

For $h > 0$, straight forward calculation (cf. Appendix A.1) shows

$$H_{t+h} = (1 - t - h)F_1 + (t + h)\Delta_x = \left(1 - \frac{h}{1-t}\right) \cdot H_t + \frac{h}{1-t} \cdot \Delta_x.$$

The Gâteaux-derivative of T at F_1 can be gained as the ordinary derivative of $T(H_t)$ at $t = 0$, compare (2.14). Recall that

$$\int_{\mathbb{R}} IF(s; H_t, T)\,dH_t(s) = E(IF(X; H_t, T)) = 0, \tag{2.25}$$

provided the denominator of (2.22) is non-zero. Then we get

$$
\begin{aligned}
\frac{d}{dt}T(H_t) &= \lim_{h \to 0} \frac{T(H_{t+h}) - T(H_t)}{h} \\
&= \lim_{h \to 0} \frac{T\left(\left(1 - \frac{h}{1-t}\right) \cdot H_t + \frac{h}{1-t} \cdot \Delta_x\right) - T(H_t)}{h} \\
&= \lim_{s \to 0} \frac{T\left((1 - s)H_t + s\Delta_x\right) - T(H_t)}{s(1 - t)} \qquad \text{due to } s = \frac{h}{1-t} \\
&= \frac{1}{1-t} \cdot IF(x; H_t, T) \\
&= \frac{1}{1-t} \int_0^\infty IF(s; H_t, T)\,d(\Delta_x - H_t)(s) \qquad \text{due to (2.25)} \\
&= \frac{1}{1-t} \int_0^\infty IF(s; H_t, T)\,d(\Delta_x - (1 - t)F_1 - t\Delta_x)(s) \\
&= \frac{1}{1-t} \int_0^\infty IF(s; H_t, T)\,d(\Delta_x - F_1)(s) - \frac{t}{1-t} \int_0^\infty IF(s; H_t, T)\,d(\Delta_x - F_1)(s)
\end{aligned}
$$

$$= \int\limits_0^\infty IF(s; H_t, T) d(\Delta_x - F_1)(s).$$

Since $T(\Delta_x) - T(F_1) = \int_0^1 \frac{d}{dt} T((1-t)F_1 + t\Delta_x) dt = \int_0^1 \frac{d}{dt} T(H_t) dt$, it follows that

$$T(\Delta_x) - T(F_1) = \int\limits_0^1 \int\limits_0^\infty IF\big(s; (1-t)F_1 + t\Delta_x, T\big) d(\Delta_x - F_1)(s) dt,$$

cf. Huber [Huber 1981], p. 38.

To derive $\mathrm{Var}(T_n)$ we assume T to be Fréchet-differentiable as well. The Taylor expansion of $\bar{H}(t) = T((1-t)F_1 + tF_n)$, $F_1 \in \mathcal{F}_\varepsilon$ in $t = 0$ is

$$\bar{H}(1) = \bar{H}(0) + \bar{H}'(0) + \tfrac{1}{2}\bar{H}''(\xi), \qquad \xi \in (0,1).$$

Because T is Fréchet-differentiable and thus Gâteaux-differentiable we have $\bar{H}'(0) = T'_{F_n - F_1}(F_1)$. Since the empirical distribution function $F_n(x; X_1, \ldots, X_n)$ can be expressed as $F_n(x; X_1, \ldots, X_n) = \frac{1}{n}\sum_{i=1}^n \Delta_{X_i}$, by the linearity of T we get

$$T'_{F_n - F_1}(F_1) = \lim_{t \to 0} \frac{T((1-t) \cdot F_1 + t \cdot \frac{1}{n}\sum_{i=1}^n \Delta_{X_i})}{t} = \frac{1}{n}\sum_{i=1}^n \lim_{t \to 0} \frac{T((1-t) \cdot F_1 + t \cdot \Delta_{X_i})}{t}$$

$$= \frac{1}{n}\sum_{i=1}^n T'_{\Delta_{X_i}}(F_1) = \frac{1}{n}\sum_{i=1}^n IF(X_i; F_1, T).$$

Furthermore

$$\frac{1}{2}\bar{H}''(\xi) = \frac{d^2}{dt^2} T((1-t)F_1 + tF_n)\big|_{t=\xi} = T''_{F_n - F_1}((1-\xi)F_1 + \xi F_n)(F_n - F_1)^2.$$

By virtue of T being a linear operator, we know that $T = T'_{F_n - F_1} = T''_{F_n - F_1}$. Thus

$$\frac{1}{2}\bar{H}''(\xi) = T((1-\xi)F_1 + \xi F_n)(F_n - F_1)^2.$$

From Fréchet differentiability it follows

$$T(F_n) - T(F_1) - T'_{F_n - F_1}(F_1)(F_n - F_1) - T''_{F_n - F_1}((1-\xi)F_1 + \xi F_n)(F_n - F_1)^2$$
$$= o_{\mathrm{P}}(F_n - F_1) \longrightarrow 0,$$

where $F = o_{\mathrm{P}}(G)$ means

$$\lim_{x \to \infty} \frac{F(x)}{G(x)} \overset{P}{=} 0.$$

To derive the order of convergence note that

$$\sqrt{n}(T(F_n) - T(F_1)) = \sqrt{n}(\bar{H}(1) - \bar{H}(0))$$

$$= \sqrt{n}\left(\bar{H}'(0) + \frac{1}{2}\bar{H}''(\xi)\right), \qquad \xi \in (0,1)$$

$$= \sqrt{n}\left(T'_{F_n - F_1}(F_1) + \frac{1}{2}T''_{F_n - F_1}((1-\xi)F_1 + \xi F_n)(F_n - F_1)^2\right)$$

$$= \sqrt{n} \cdot \frac{1}{n}\sum_{i=1}^{n} IF(X_i; F_1, T)$$

$$+ \sqrt{n} \cdot \frac{1}{2}T''_{F_n - F_1}((1-\xi)F_1 + \xi F_n) \cdot (F_n - F_1)^2$$

$$= \sqrt{n}\frac{1}{n}\sum_{i=1}^{n} IF(X_i; F_1, T) + \sqrt{n} \cdot \frac{1}{2}o_P(F_n - F_1).$$

If $\sqrt{n}||F_n - F_1|| = o(1)$ we get

$$\sqrt{n}(T(F_n) - T(F_1)) = \frac{1}{\sqrt{n}}\sum_{i=1}^{n} IF(X_i; F_1, T) + o(1).$$

Assume the random sample $X_1, \ldots, X_n$ to be i.i.d. with finite variance $\sigma^2 > 0$, then by the Central Limit Theorem and (2.22)

$$\sqrt{n} \cdot \frac{1}{n}\sum_{i=1}^{n} IF(X_i; F_1, T(F_1)) \xrightarrow{\mathcal{L}} N(0, \sigma^2).$$

Thus $\sqrt{n}(T(F_n) - T(F_1))$ and therefore T_n have asymptotic variance

$$\sigma^2_{\text{asym}} = \text{Var}\left(IF(X; F_1, T)\right) \stackrel{(2.23)}{=} E\left(IF^2(X; F_1, T)\right). \tag{2.26}$$

It is important to realise that (2.26) needs $\sqrt{n}||F_n - F_1|| = o(1)$. If this condition does not hold or cannot be proved, the above approach is not applicable.

2.4 Defining Robust M-Estimators Using the Influence Function

One of the main differences in the concepts of qualitative (cf. Section 1.3.1) and quantitative robustness (cf. Section 1.3.2 and 2.3) are their different perspectives on statistical problems. Qualitative robustness follows a global approach whereas quantitative robustness looks from a local point of view.

As Definition 1.3 says, a robust sequence of estimators $(T_n)_{n\in\mathbb{N}}$ is defined by a functional T that is continuous in the weak*-topology.
In contrast, quantitative robustness (Section 2.3) favours a more manageable but local

approach. It turns out, that the influence function (2.15) can be used to find a robust M-estimator in the quantitative sense, cf. Hampel [Hampel 1974].

Now, Hampel [Hampel 1974] looks for M-estimators with asymptotic minimum variance given that

$$\int_{\mathbb{R}} \psi(s; \theta) dF_\theta(s) = 0$$

and $\gamma = \sup_x |IF(x; F_\theta, T)| < k, \qquad k > 0.$

The following lemma was first stated by Hampel [Hampel 1974]. It is also mentioned by Huber [Huber 1981] and Hamepl et al. [Hampel et al. 1986]. Because we are going to use the result of the lemma, we will work out the proof more explicitly than it is done by Hampel et al. [Hampel et al. 1986].

Lemma 2.10
Let the parameter space $\Theta \subset \mathbb{R}$ be an open, convex set and $\theta_0 \in \Theta$. $\mathcal{F}_\theta$ denotes the set of all continuous distribution functions F_θ on the sample space $(\Omega', \mathcal{A}') = (\mathbb{R}, \mathfrak{B})$. Assume that for all $x \in \mathbb{R}$

$$\frac{\frac{\partial}{\partial \theta} f_\theta(x)\big|_{\theta=\theta_0}}{f_{\theta_0}(x)} = \frac{\partial}{\partial \theta} \ln f_\theta(x)\big|_{\theta=\theta_0} \quad \text{exists and} \quad \int_{\mathbb{R}} \frac{\partial}{\partial \theta} \ln f_\theta(s)\big|_{\theta=\theta_0} dF_{\theta_0}(s) = 0. \qquad (2.27)$$

In addition, the Fisher information

$$I(\theta_0) = \int_{\mathbb{R}} \left(\frac{\partial}{\partial \theta} \ln f_\theta(s)\big|_{\theta=\theta_0} \right)^2 dF_{\theta_0}(s)$$

is assumed to be finite and greater than 0.
Let $b > 0$ be some constant. Then there exists a real number $a \in \mathbb{R}$ such that for θ_0 and

$$\bar\psi(x; \theta_0) = \max\left\{ -b, \min\left\{ g(x; \theta_0) - a, b \right\} \right\}, \qquad g(x; \theta) = \frac{\partial}{\partial \theta} \ln f_\theta(x) \qquad (2.28)$$

the following hold

$$0 < \int_{\mathbb{R}} \bar\psi(x; \theta) \frac{\partial}{\partial \theta} \ln f_\theta(s)\big|_{\theta=\theta_0} dF_{\theta_0}(s) =: d \qquad (2.29)$$

$$\frac{\int_{\mathbb{R}} \bar\psi^2(s; \theta_0) dF_{\theta_0}(s)}{\left[\int_{\mathbb{R}} \bar\psi(s; \theta) \frac{\partial}{\partial \theta} \ln f_\theta(s)\big|_{\theta=\theta_0} dF_{\theta_0}(s) \right]^2} \qquad (2.30)$$

$$= \min_{\psi \in \Psi} \left\{ \frac{\int_{\mathbb{R}} \psi^2(s; \theta_0) dF_{\theta_0}(s)}{\left[\int_{\mathbb{R}} \psi(s; \theta) \frac{\partial}{\partial \theta} \ln f_\theta(s)\big|_{\theta=\theta_0} dF_{\theta_0}(s) \right]^2} \right\},$$

where Ψ contains all functions $\psi : \mathbb{R} \times \Theta \to \mathbb{R}$ satisfying

$$\int_{\mathbb{R}} \psi(s;\theta) dF_\theta(s) = 0 \tag{2.31}$$

$$\int_{\mathbb{R}} \psi(s;\theta) \frac{\partial}{\partial \theta} \ln f_\theta(s)\big|_{\theta=\theta_0} dF_{\theta_0}(s) \neq 0 \tag{2.32}$$

$$\sup_{x \in \mathbb{R}} \left| \frac{\psi(x;\theta)}{\int_{\mathbb{R}} \psi(s;\theta) \frac{\partial}{\partial \theta} \ln f_\theta(s)\big|_{\theta=\theta_0} dF_{\theta_0}(s)} \right| \leq \frac{b}{d} \tag{2.33}$$

Before we prove the lemma, we have a closer look at the requirements. If (2.27) hold, then

$$I(\theta_0) = \int_{\mathbb{R}} \frac{\partial}{\partial \theta} \ln f_\theta(s)\big|_{\theta=\theta_0} dF_{\theta_0}(s) = \mathrm{Var}\left(\frac{\partial}{\partial \theta} \ln f_\theta(X)\big|_{\theta=\theta_0} \right),$$

cf. Lehmann, Casella [Lehmann, Casella 1998], p. 116 Lemma 5.3.
Additionally, differentiating (2.31) on both sides with respect to some θ_0 yields

$$\frac{\partial}{\partial \theta} \left[\int_{\mathbb{R}} \psi(s;\theta) dF_\theta(s) \right]_{\theta=\theta_0} = \frac{\partial}{\partial \theta} \int_{\mathbb{R}} \psi(s;\theta) dF_{\theta_0}(s) + \int_{\mathbb{R}} \psi(s;\theta_0) \frac{\partial}{\partial \theta} f_\theta(s)\big|_{\theta=\theta_0} ds$$

$$= \frac{\partial}{\partial \theta} \int_{\mathbb{R}} \psi(s;\theta) dF_{\theta_0}(s) + \int_{\mathbb{R}} \psi(s;\theta_0) \frac{\partial}{\partial \theta} \ln f_\theta(s)\big|_{\theta=\theta_0} dF_{\theta_0}(s)$$

$$= 0.$$

Therefore d in (2.29) can be written as

$$d = -\frac{\partial}{\partial \theta} \int_{\mathbb{R}} \bar{\psi}(s;\theta) dF_{\theta_0}(s)\big|_{\theta=\theta_0}. \tag{2.34}$$

In the same way, the denominator of (2.33) can be substituted.
Due to (2.31) any M-estimator T_n with functional T defined by

$$\int_{\mathbb{R}} \bar{\psi}(s;T(F_\theta)) dF_\theta(s) = 0$$

is Fisher-consistent according to Section 2.2.3. Besides (2.33) gives the gross error sensitivity (2.16) for $\theta = \theta_0 = T(F_{\theta_0})$ because of (2.34).

Note that if F and T fulfil all necessary requirements and $\theta_0 = T(F_{\theta_0})$, then because of (2.19), (2.26), (2.31) and (2.34) T_n has minimal asymptotic variance σ_{asym}^2.

For better reading, we will write briefly $\frac{\partial}{\partial \theta_0} \ln f_{\theta_0}(s)$ instead of $\frac{\partial}{\partial \theta} \ln f_\theta(s)|_{\theta=\theta_0}$. Recall that θ is a scale parameter and the support of F_θ is supposed to be $\mathbb{R}^+ = (0,\infty)$.

Proof: Assume $b > 0$ and choose $\theta_0 \in \Theta$.
First we show, that the function

$$h : \mathbb{R} \to \mathbb{R}, \qquad h(a) = \int_{\mathbb{R}} \max\{-b, \min\{s - a, b\}\} dF_{\theta_0}(s)$$

has a zero.
Define $\bar{\psi}_a : \mathbb{R} \to \mathbb{R}$ to be

$$\bar{\psi}_a(x) = \max\{-b, \min\{x - a, b\}\},$$

then obviously $|\psi_a(x)| \leq b$ and $\int |\psi_a(s)| dF_{\theta_0}(s) < \infty$. Let $(a_n)_{n \in \mathbb{N}}$ denotes a sequence with $\lim_{n \to \infty} a_n = a$. Then

$$h(a_n) = \int_{\mathbb{R}} \psi_{a_n}(s) dF_{\theta_0}(s) \qquad \text{and} \qquad |\psi_{a_n}(s)| \leq b \quad \text{for all } n \in \mathbb{N}.$$

Applying the Dominated Convergence Theorem (e.g. Bauer [Bauer 1992]) we conclude

$$\lim_{n \to \infty} h(a_n) = \lim_{n \to \infty} \int_{\mathbb{R}} \psi_{a_n}(s) dF_{\theta_0}(s) = \int_{\mathbb{R}} \lim_{n \to \infty} \psi_{a_n}(s) dF_{\theta_0}(s)$$

$$= \int_{\mathbb{R}} \max\{-b, \min\{s - \lim_{n \to \infty} a_n, b\}\} dF_{\theta_0}(s) = h(\lim_{n \to \infty} a_n) = h(a).$$

It follows that h is continuous in a.
Since

$$\lim_{a \to -\infty} h(a) = \int_{\mathbb{R}} \lim_{a \to -\infty} \max\{-b, \min\{s - a, b\}\} dF_{\theta_0}(s) = b \int_{\mathbb{R}} dF_{\theta_0}(s) = b,$$

$$\lim_{a \to \infty} h(a) = \int_{\mathbb{R}} \lim_{a \to \infty} \max\{-b, \min\{s - a, b\}\} dF_{\theta_0}(s) = -b \int_{\mathbb{R}} dF_{\theta_0}(s) = -b,$$

there exists an $a \in \mathbb{R}$ with $h(a) = 0$ because of the Mean Value Theorem.
So we have shown that there exists a zero of h.

Next we prove $d > 0$.
Define $\bar{\psi} : \mathbb{R} \times \Theta \to \mathbb{R}$ to be

$$\bar{\psi}(x, \theta) = \max\{-b, \min\{\tfrac{\partial}{\partial \theta} \ln f_\theta(s) - a, b\}\}$$
$$= \max\{a - b, \min\{\tfrac{\partial}{\partial \theta} \ln f_\theta(s), a + b\}\} - a.$$

and

$$d = \int_{\mathbb{R}} \bar{\psi}(s; \theta_0) \tfrac{\partial}{\partial \theta_0} \ln f_{\theta_0}(s) \, dF_{\theta_0}(s)$$

$$= \int_{\mathbb{R}} \left(\max \left\{ a - b, \min \left\{ \tfrac{\partial}{\partial \theta_0} \ln f_{\theta_0}(s), a + b \right\} \right\} - a \right) \cdot \tfrac{\partial}{\partial \theta_0} \ln f_{\theta_0}(s) \, dF_{\theta_0}(s)$$

$$= \int_{\mathbb{R}} \max \left\{ a - b, \min \left\{ \tfrac{\partial}{\partial \theta_0} \ln f_{\theta_0}(s), a + b \right\} \right\} \cdot \tfrac{\partial}{\partial \theta_0} \ln f_{\theta_0}(s) \, dF_{\theta_0}(s).$$

There are three possible cases for the relation of a and b, $|a| \leq b$, $a > b$ and $a < -b$. We start with the case $|a| \leq b$.
Put

$$M_1 = \left\{ s \in \mathbb{R}_0^+ : \tfrac{\partial}{\partial \theta_0} \ln f_{\theta_0}(s) > 0 \right\} \qquad \text{and} \qquad M_2 = \left\{ s \in \mathbb{R}_0^+ : \tfrac{\partial}{\partial \theta_0} \ln f_{\theta_0}(s) < 0 \right\},$$

then it follows

$$d = \int_{M_1} \min \left\{ \tfrac{\partial}{\partial \theta_0} \ln f_{\theta_0}(s), a + b \right\} \cdot \tfrac{\partial}{\partial \theta_0} \ln f_{\theta_0}(s) dF_\theta(x)$$

$$+ \int_{M_2} \max \left\{ a - b, \tfrac{\partial}{\partial \theta} \ln f_\theta(s) \right\} \cdot \tfrac{\partial}{\partial \theta_0} \ln f_{\theta_0}(s) dF_{\theta_0}(x).$$

By assumption $a + b \geq 0$, and thus the minimum in the first part of the sum is non-negative. Besides $a - b \leq 0$ and on M_2 we have $\tfrac{\partial}{\partial \theta_0} \ln f_{\theta_0}(s) < 0$. Therefore the second part of the sum is non-negative as well, so is d.

We will now show that even $d > 0$. Assume $d = 0$, then both integrals have to be zero because of their non-negativity. But only one term of $a - b$ and $a + b$ can be zero at a time, meaning M_1 and M_2 have to be null sets. This results in a Fisher Information $I(\theta_0) = 0$ since

$$I(\theta_0) = \int_{\mathbb{R}} \left(\tfrac{\partial}{\partial \theta_0} \ln f_{\theta_0}(s) \right)^2 dF_{\theta_0}(s)$$

$$= \int_{M_1} \left(\tfrac{\partial}{\partial \theta_0} \ln f_{\theta_0}(s) \right)^2 dF_{\theta_0}(s) + \int_{M_2} \left(\tfrac{\partial}{\partial \theta_0} \ln f_{\theta_0}(s) \right)^2 dF_{\theta_0}(s)$$

$$= 0,$$

which contradicts the assumption.

We now turn to the case $a > b$ yielding

$$\bar{\psi}(x; \theta_0) = \max \left\{ -b, \min \left\{ \tfrac{\partial}{\partial \theta_0} \ln f_{\theta_0}(x) - a, b \right\} \right\}$$

$$= \max \left\{ 0, \min \left\{ \tfrac{\partial}{\partial \theta_0} \ln f_{\theta_0}(x) + b - a, 2b \right\} \right\} - b,$$

that means $\bar{\psi}(x; \theta_0) + b \geq 0$. Applying (2.27) as well as (2.32) gives

$$
\begin{aligned}
d &= \int_{\mathbb{R}} \bar{\psi}(s; \theta_0) \frac{\partial}{\partial \theta_0} \ln f_{\theta_0}(s) \, dF_{\theta_0}(s) \\
&= \int_{\mathbb{R}} \max \left\{ 0, \min \left\{ \frac{\partial}{\partial \theta_0} \ln f_{\theta_0}(s) + b - a, 2b \right\} \right\} dF_{\theta_0}(s) > 0.
\end{aligned}
$$

The last case deals with $a < -b$, that is

$$
\begin{aligned}
\bar{\psi}(x; \theta_0) &= \max \left\{ -b, \min \left\{ \frac{\partial}{\partial \theta_0} \ln f_{\theta_0}(x) - a, b \right\} \right\} \\
&= \max \left\{ -2b, \min \left\{ \frac{\partial}{\partial \theta_0} \ln f_{\theta_0}(x) - (a + b), 0 \right\} \right\} + b
\end{aligned}
$$

meaning

$$
\bar{\psi}(x; \theta) - b \geq 0.
$$

Again we use (2.27) and (2.32) to get

$$
\begin{aligned}
d &= \int_{\mathbb{R}} \bar{\psi}(s; \theta_0) \frac{\partial}{\partial \theta_0} \ln f_{\theta_0}(s) \, dF_{\theta_0}(s) \\
&= \int_{\mathbb{R}} \max \left\{ -2b, \min \left\{ \frac{\partial}{\partial \theta_0} \ln f_{\theta_0}(s) - (a + b), 0 \right\} \right\} dF_{\theta_0}(s) > 0.
\end{aligned}
$$

We are left proving that $\bar{\psi}(x; \theta_0) = \max \left\{ -b, \min \left\{ \frac{\partial}{\partial \theta_0} \ln f_{\theta_0}(s) - a \right\} \right\}$ minimises the function

$$
\frac{\int_{\mathbb{R}} \psi^2(s; \theta) dF_{\theta}(s)}{\left[\int_{\mathbb{R}} \psi(s; \theta_0) \frac{\partial}{\partial \theta_0} \ln f_{\theta_0}(s) dF_{\theta_0}(s) \right]^2}. \tag{2.35}
$$

Without loss of generality for ψ we can assume

$$
\int_{\mathbb{R}} \psi(s; \theta_0) \frac{\partial}{\partial \theta_0} \ln f_{\theta_0}(s) dF_{\theta_0}(s) = d,
$$

because for some function $\hat{\psi}$ fulfilling conditions (2.29) to (2.33), too, we get

$$
0 < \int_{\mathbb{R}} \hat{\psi}(s; \theta_0) \frac{\partial}{\partial \theta_0} \ln f_{\theta_0}(s) \, dF_{\theta_0}(s) = e \neq d
$$

and can consider $\frac{d}{e} \hat{\psi}$ instead.

In other words the problem of minimising (2.35) reduces to minimising $\int \psi^2(s;\theta_0)dF_{\theta_0}(s)$. We have

$$\int_{\mathbb{R}} \left(\tfrac{\partial}{\partial\theta_0} \ln f_{\theta_0}(s) - a - \psi(s;\theta_0) \right)^2 dF_{\theta_0}(s)$$

$$= \int_{\mathbb{R}} \left(\tfrac{\partial}{\partial\theta_0} \ln f_{\theta_0}(s) - a \right)^2 dF_{\theta_0}(s) - 2 \int_{\mathbb{R}} \left(\tfrac{\partial}{\partial\theta_0} \ln f_{\theta_0}(s) - a \right) \psi(s;\theta_0)dF_{\theta_0}(s)$$

$$+ \int_{\mathbb{R}} \psi^2(s;\theta_0)dF_{\theta_0}(s)$$

$$= \int_{\mathbb{R}} \left(\tfrac{\partial}{\partial\theta_0} \ln f_{\theta_0}(s) - a \right)^2 dF_{\theta_0}(s) - 2d + \int_{\mathbb{R}} \psi^2(s;\theta_0)dF_{\theta_0}(s)$$

because of (2.31) and (2.34). Obviously, a function ψ minimising $\int \psi^2(s;\theta_0)dF_{\theta_0}(s)$ also minimises $\int(\tfrac{\partial}{\partial\theta_0} \ln f_{\theta_0}(s) - a - \psi(s;\theta_0))^2 dF_{\theta_0}(s)$. Splitting the integration domain of the first integral given above, we get

$$\int_{\mathbb{R}} \left(\tfrac{\partial}{\partial\theta_0} \ln f_{\theta_0}(s) - a - \psi(s;\theta_0) \right)^2 dF_{\theta_0}(s)$$

$$= \int_{\left\{\tfrac{\partial}{\partial\theta_0} \ln f_{\theta_0}(s)-a<-b\right\}} \left(\tfrac{\partial}{\partial\theta_0} \ln f_{\theta_0}(s) - a - \psi(s;\theta_0) \right)^2 dF_{\theta_0}(s)$$

$$+ \int_{\left\{-b\le\tfrac{\partial}{\partial\theta_0} \ln f_{\theta_0}(s)-a\le b\right\}} \left(\tfrac{\partial}{\partial\theta_0} \ln f_{\theta_0}(s) - a - \psi(s;\theta_0) \right)^2 dF_{\theta_0}(s)$$

$$+ \int_{\left\{\tfrac{\partial}{\partial\theta_0} \ln f_{\theta_0}(s)-a>b\right\}} \left(\tfrac{\partial}{\partial\theta_0} \ln f_{\theta_0}(s) - a - \psi(s;\theta_0)^2 \right) dF_{\theta_0}(s).$$

Recalling that

$$\sup_{x\in\mathbb{R}} \left| \frac{\psi(x;\theta)}{\int_{\mathbb{R}} \psi(s;\theta)\tfrac{\partial}{\partial\theta_0} \ln f_{\theta_0}(s)\,dF_{\theta_0}(s)} \right| = \sup_{x\in\mathbb{R}} \frac{|\psi(x;\theta)|}{d} \le \frac{b}{d},$$

i.e. $|\psi(x;\theta)| \le b$, we get for the optimal $\bar\psi$

$$\bar\psi(x;\theta) = \begin{cases} -b, & \tfrac{\partial}{\partial\theta_0} \ln f_{\theta_0}(s) - a < -b \\ \tfrac{\partial}{\partial\theta_0} \ln f_{\theta_0}(s) - a, & -b < \tfrac{\partial}{\partial\theta_0} \ln f_{\theta_0}(s) - a < b \\ b, & \tfrac{\partial}{\partial\theta_0} \ln f_{\theta_0}(s) - a > b \end{cases}$$

$$= \max\left\{ -b, \min\left\{ \tfrac{\partial}{\partial\theta_0} \ln f_{\theta_0}(s) - a, b \right\} \right\}$$

$\square$

With the example in Section 2.2 it is apparent that the M-estimator T_n with functional T solving

$$\int_{\mathbb{R}} \bar{\psi}(s; T(F_\theta)) dF_\theta(s) = 0$$

is a maximum-likelihood estimator with limited range.

Our next step is going to be the determination of a robust M-estimator for the data model from Section 1.4.

Chapter 3

Robust Estimators in the Γ-Model

This chapter is going to combine our results from the two preceding chapters.
Recall our data model (1.10) from Section 1.4. We assume our claim data to come from a distribution that in a major part is equal to a $\Gamma(\alpha, \theta)$-distribution. For future reference, (1.10) is briefly denoted by Γ-*model*.

In Lemma 2.10 we learned that the function ψ for M-estimators minimising the asymptotic variance (2.26) is of the form (2.28)

$$\psi(x;\theta) = \max\left\{-b, \min\left\{\frac{\partial}{\partial\theta}\ln f_\theta(x) - a, b\right\}\right\}, \quad b > 0, \ a = a(b) \in \mathbb{R}.$$

What we do next is to compute ψ for our Γ-model. Besides, this chapter presents our main results. We introduce the M-estimator T_n defined by the function ψ given above. By discussing the consistency of T_n, we develop a second robust M-estimator $\tilde{T}_n$.
In Sections 3.3.1 and 3.3.2, we will explain the ideas of computing optimal values of a in ψ and $\tilde{\psi}$ as well as prove the results.

3.1 The M-Estimator in the Γ-Model

As we already stated in Section 1.4, the parameter to be estimated in our model is $\theta \in \Theta = (0, \infty)$, the scale parameter of the $\Gamma(\alpha, \theta)$-distribution. It is thus not unexpected that we are looking for a scale M-estimator as described in Section 2.2.2.

Let F_θ be a continuous distribution function with scale parameter θ and density f_θ. Then

$$\frac{\partial}{\partial\theta}\ln f_\theta(x) = \frac{\frac{\partial}{\partial\theta}f_\theta(x)}{f_\theta(x)} = \frac{\frac{\partial}{\partial\theta}\left(\frac{1}{\theta}f_1\left(\frac{x}{\theta}\right)\right)}{\frac{1}{\theta}f_1\left(\frac{x}{\theta}\right)} = -\frac{1}{\theta} - \frac{x}{\theta^2}\cdot\frac{f_1'\left(\frac{x}{\theta}\right)}{f_1\left(\frac{x}{\theta}\right)} \tag{3.1}$$

and according to (2.28) we get

$$\psi(x;\theta) = \max\left\{-b, \min\left\{-\frac{1}{\theta} - \frac{x}{\theta^2}\cdot\frac{f_1'\left(\frac{x}{\theta}\right)}{f_1\left(\frac{x}{\theta}\right)} - a, b\right\}\right\}.$$

In Section 2.2.2 we also explained that a scale M-estimator is sufficiently described by the function $\hat{\psi} : \mathbb{R} \to \mathbb{R}$ with

$$\hat{\psi}\left(\frac{x}{\theta}\right) = \psi(x; \theta).$$

For the Γ-model this means

$$-\frac{1}{\theta} - \frac{x}{\theta^2} \cdot \frac{f_1'\left(\frac{x}{\theta}\right)}{f_1\left(\frac{x}{\theta}\right)}\Bigg|_{\theta=1} = -\frac{1}{\theta} - \frac{x}{\theta^2} \cdot \left(\frac{\alpha-1}{\frac{x}{\theta}} - 1\right)\Bigg|_{\theta=1} = x - \alpha$$

and therefore

$$\hat{\psi}(x) = \max\left\{-b, \min\left\{x - (\alpha + a), b\right\}\right\}. \tag{3.2}$$

Remark 3.1
We have not closely examined that all requirements of Lemma 2.10 are fulfilled but this is easy to see.
Certainly, $\Theta = (0, \infty) \subset \mathbb{R}$ is an open and convex set. Furthermore, because $f_\theta(x) > 0$

$$\int_0^\infty \frac{\partial}{\partial \theta} \ln f_\theta(s) dF_\theta(s) = \int_0^\infty \frac{\partial}{\partial \theta} f_\theta(s) ds = \frac{\partial}{\partial \theta} \int_0^\infty f_\theta(s) ds = 0,$$

where the exchange of differentiation and integration is possible according to Mangolft, Knopp [Mangoldt, Knopp 1975], Theorem on p. 335.
In Lehmann, Casella [Lehmann, Casella 1998], p. 117 Example 5.5, Fisher information for the different parameters of the Γ-distribution is calculated. Hence $I(\theta)$ exists.

Remark 3.2
Recall that the calculation of the asymptotic variance σ_{asym}^2 according to formula (2.26) requires the validation of $\sqrt{n}||F_n - F_1|| = o(1)$. We have not shown yet this to be true for the $\Gamma(\alpha, 1)$-distribution function. But in Chapter 4 we are going to prove that our robust M-estimator is asymptotically normal distributed with variance given as in (2.26).

Our scale M-estimator $T_{n,i}$ for risk parameter θ_i is now defined through

$$\sum_{j=1}^n \psi\left(\frac{X_{ij}}{T_{n,i}}\right) = 0, \quad \psi(x) = \max\left\{-b, \min\left\{x - (\alpha + a), b\right\}\right\} \tag{3.3}$$

and if $T_{n,i}$ is consistent, then

$$\int_0^\infty \max\left\{-b, \min\left\{\frac{s}{T(F_1)} - (\alpha + a), b\right\}\right\} dF_1(s) = 0, \quad T(F_\theta) = \theta \cdot T(F_1).$$

Because from now on we are only dealing with scale M-estimators, we rename $\hat{\psi}$ in ψ.

So far our scale M-estimator $T_{n,i}$ is only given implicitly in equation (3.3), which is quite uncomfortable to work with. But we will see later that we can find a more manageable estimator $\tilde{T}_{n,i}$.

3.2   The Estimator $T_{n,i}$

Let $X_{i1}, \ldots, X_{in}$ be a random sample drawn from $F \in \mathcal{F}_\varepsilon(F_{\theta_i})$ with

$$\mathcal{F}_\varepsilon(F_{\theta_i}) = \{F : F = (1 - \varepsilon) \cdot \Gamma(\alpha, \theta_i) + \varepsilon \cdot Par(\lambda, x_0)\}, \qquad 0 < \varepsilon < 1,$$

then our scale estimator $T_{n,i}$ for θ_i is defined by

$$\sum_{j=1}^{n} \psi \left(\frac{X_{ij}}{T_{n,i}} \right) = 0, \qquad \psi(x) = \max\{-b, \min\{x - (\alpha + a), b\}\}. \tag{3.4}$$

Now $l_1, l_2 \in \mathbb{N}$ denote the indices such that

$$\begin{aligned}
\frac{X_{i(l_1)}}{T_{n,i}} &< \alpha + a - b, & \frac{X_{i(l_1+1)}}{T_{n,i}} &\geq \alpha + a - b, \\
\frac{X_{i(n-l_2)}}{T_{n,i}} &\leq \alpha + a + b, & \frac{X_{i(n-l_2+1)}}{T_{n,i}} &> \alpha + a + b,
\end{aligned} \tag{3.5}$$

where $X_{i(1)} \leq X_{i(2)} \leq \ldots \leq X_{i(n)}$ is the ordered sample.

To derive our M-estimator $T_{n,i}$, we go back to the classical definition of an estimator being a real function, that computes for all samples $x_{i1}, \ldots, x_{in}$ of $X_{i1}, \ldots, X_{in}$ an estimate of the parameter under consideration.

This means, the M-estimate $T_{n,i}(\mathbf{x})$ of sample $\mathbf{x} = (x_{i1}, \ldots, x_{in})$ solves

$$\sum_{j=1}^{n} \psi \left(\frac{x_{ij}}{T_{n,i}(\mathbf{x})} \right) = 0,$$

which can be rewritten as

$$\begin{aligned}
\sum_{j=1}^{n} \psi \left(\frac{x_{ij}}{T_{n,i}(\mathbf{x})} \right) &= \sum_{j=1}^{n} \psi \left(\frac{x_{i(j)}}{T_n(\mathbf{x})} \right) \\
&= \sum_{j=1}^{n} \max \left\{ -b, \min \left\{ \frac{x_{i(j)}}{T_n(\mathbf{x})} - (\alpha + a), b \right\} \right\} \\
&= -l_1 b + \sum_{i=l_1+1}^{n-l_2} \left(\frac{x_{i(j)}}{T_n(\mathbf{x})} - (\alpha + a) \right) + l_2 b \\
&= b(l_2 - l_1) - (n - l_2 - l_1)(\alpha + a) + \sum_{j=l_1+1}^{n-l_2} \frac{x_{i(j)}}{T_n(\mathbf{x})} \\
&\stackrel{!}{=} 0,
\end{aligned}$$

where $r_{i(1)} \leq r_{i(2)} \leq r_{i(n)}$

It follows that

$$\begin{aligned}
T_{n,i}(\mathbf{x}) &= \frac{1}{(n - l_2 - l_1)(\alpha + a) + l_1 b - l_2 b} \cdot \sum_{j=l_1+1}^{n-l_2} x_{i(j)} \\
&= \frac{1}{\alpha + a - \frac{l_1}{n}(\alpha + a - b) - \frac{l_2}{n}(\alpha + a + b)} \cdot \frac{1}{n} \sum_{j=l_1+1}^{n-l_2} x_{i(j)},
\end{aligned}$$

and our estimator $T_{n,i}$ will be

$$T_{n,i} = \frac{1}{\alpha + a - \frac{l_1}{n}(\alpha + a - b) - \frac{l_2}{n}(\alpha + a + b)} \cdot \frac{1}{n} \sum_{j=l_1+1}^{n-l_2} X_{i(j)} \tag{3.6}$$

In Gisler, Reinhard [Gisler, Reinhard 1993] the function ψ is defined as follows

$$\psi(x) = \min\{x - 1, 1\}. \tag{3.7}$$

They also consider a Γ-model which they motivate by analysing the shape of the influence function of the exact Bayesian estimator. Besides they consider loss ratios instead of claims amounts.

Choosing $b = 1$ and a such that $\alpha + a = 1$, i.e. $a = b - \alpha$, equation (3.7) is a special case of (3.2).

3.2.1   Consistency of the Estimator $T_{n,i}$

An estimator δ_n of a parameter $\theta \in \Theta$ is called *consistent*, if

$$\delta_n \xrightarrow{P} \theta, \qquad \text{for all } \theta \in \Theta,$$

cf. Lehmann, Casella [Lehmann, Casella 1998].

Since $T_{n,i}$ in (3.4) is a scale estimator, we can apply Thall's consistency theorem in Thall [Thall 1979], Theorem 1. Thall uses Huber's Lemma to show the consistency of a location M-estimator (cf. Huber [Huber 1964], Lemma 3) and adjusts the conditions towards a scale M-estimator.

We use the fact that, if θ is a scale parameter of a distribution function F_θ of some random variable X, then $\mu = \ln \theta$ is a location parameter of the distribution G_μ of the random variable $Y = \ln X$ (cf. Appendix A.2).

From (3.1) we know that for a scale parameter θ

$$\frac{\partial}{\partial \theta} \ln f_\theta(x) = -\frac{1}{\theta} - \frac{x}{\theta^2} \cdot \frac{f_1'\left(\frac{x}{\theta}\right)}{f_1\left(\frac{x}{\theta}\right)}.$$

Thus for $x = e^y$

$$\begin{aligned} f_\theta(x) &= \frac{d}{dx} F_\theta(x) \\ &= \frac{d}{dx} G_\mu(\ln x) = \frac{1}{e^y} g_\mu(y) = \frac{1}{e^y} g_{\ln \theta}(y) \end{aligned}$$

and

$$\frac{\partial}{\partial\theta}\ln f_\theta(x) = \frac{\partial}{\partial\theta}\ln\left(\frac{1}{e^y}g_{\ln\theta}(y)\right) = \frac{1}{g_{\ln\theta}(y)}\cdot\frac{\partial}{\partial\theta}g_{\ln\theta}(y)$$

$$= \frac{\frac{\partial}{\partial\theta}g_0(y-\ln\theta)}{g_0(y-\ln\theta)} = -\frac{1}{\theta}\cdot\frac{g_0'(y-\ln\theta)}{g_0(y-\ln\theta)}$$

$$= -e^{-\mu}\cdot\frac{g_0'(y-\mu)}{g_0(y-\mu)} = e^{-\mu}\cdot\frac{\partial}{\partial\mu}\ln g_\mu(y).$$

As we already stated a couple of times, in the case of location and scale M-estimators, we only need to determine ψ_{loc}, ψ_{scale} for $\theta_{\mathrm{loc}}=0$ or $\theta_{\mathrm{scale}}=1$ and then apply $\psi_{\mathrm{loc}}(x)=\hat\psi(x-\theta)=\psi(x;\theta)$ and $\psi_{\mathrm{scale}}(x)=\hat\psi(x/\theta)=\psi(x;\theta)$, respectively.
Hence based on the above calculation, we get for the relationship between ψ_{loc} and ψ_{scale}

$$\psi_{\mathrm{scale}}(x) = \psi_{\mathrm{loc}}(\ln x) \quad\text{and}\quad \psi_{\mathrm{scale}}(e^y) = \psi_{\mathrm{loc}}(y), \tag{3.8}$$

respectively.
Now let T_{scale} be a scale equivariant functional defined by

$$\int_{\mathbb{R}} \psi_{\mathrm{scale}}\left(\frac{x}{T_{\mathrm{scale}}(F_1)}\right) dF_1(x) = 0$$

and T_{loc} a location equivariant functional defined by

$$\int_{\mathbb{R}} \psi_{\mathrm{loc}}\left(y - T_{\mathrm{loc}}(G_0)\right) dG_0(y) = 0.$$

Assume that both zeros are unique, then because of

$$\int_{\mathbb{R}} \psi_{\mathrm{scale}}\left(\frac{s}{T_{\mathrm{scale}}(F_1)}\right) dF_1(s) = \int_{\mathbb{R}} \psi_{\mathrm{loc}}\left(\ln s - \ln T_{\mathrm{scale}}(F_1)\right) dF_1(s)$$

$$= \int_{\mathbb{R}} \psi_{\mathrm{loc}}\left(t - \ln T_{\mathrm{scale}}(F_1)\right) dG_0(t)$$

we get

$$T_{\mathrm{loc}}(G_0) = \ln T_{\mathrm{scale}}(F_1).$$

This one-to-one transformation enables us to transfer conclusions from a location M-estimator to a scale M-estimator.

Because Thall's Theorem for the consistency of a scale M-estimator is based on Huber's Lemma of consistency of a location M-estimator, we state here both for convenience.

Lemma 3.3 ([Huber 1964], Lemma 3)
Define $\lambda(\theta) = \int_{\mathbb{R}} \psi_{loc}(t-\theta)dF_\theta(t)$.
Assume that there is a value c such that $\lambda(\theta) > 0$ for $\theta < c$ and $\lambda(\theta) < 0$ for $\theta > c$. Then $T_n \longrightarrow c$ a.s. and in probability.

Theorem 3.4 ([Thall 1979], Theorem 1)

Let

$$\lambda(\theta) = \int\limits_0^\infty \psi\left(\tfrac{t}{\theta}\right) dF_\theta(t), \; \theta > 0.$$

If there is a value $c > 0$ such that $\lambda(\theta) > 0$ for $\theta < c$ and $\lambda(\theta) < 0$ for $\theta > c$, then $T_n \longrightarrow c$ a.s.

We now turn our attention to the assumptions Huber uses in his proof for location M-estimators.

First, he assumes the function ρ_{loc} in equation (2.2)

$$T_{n,i} = \underset{\theta \in \Theta}{\mathrm{argmin}} \sum_{j=1}^n \rho_{\mathrm{loc}}(X_{ij} - \theta)$$

to be defined on $\mathbb{R}$, real-valued, continuous and convex. Furthermore $\rho_{\mathrm{loc}}(x) \to \infty$ for $x \to \pm\infty$.

Defining

$$\rho_{\mathrm{scale}} : \mathbb{R} \to \mathbb{R}, \qquad \rho_{\mathrm{scale}}(x) = \begin{cases} -bx - \tfrac{1}{2}(\alpha + a - b)^2 & 0 \le x < \alpha + a - b \\ \tfrac{1}{2}x^2 - (\alpha + a)x & \alpha + a - b \le x \le \alpha + a + b \\ bx - \tfrac{1}{2}(\alpha + a + b)^2 & x > \alpha + a + b \end{cases},$$

we get a function ρ_{scale} accomplishing the requirements.

The differentiablity of ρ_{scale} is shown in Appendix A.3.

Note that $\psi_{\mathrm{scale}}(x) = (d/dx)\rho_{\mathrm{scale}}(x)$ is a monotonically increasing function being smaller than 0 for x close to zero and greater than zero for large values of x.

In his lemma, Huber uses the function

$$\lambda_{\mathrm{loc}}(\theta_{\mathrm{loc}}) = E\left(\psi_{\mathrm{loc}}(X - \theta_{\mathrm{loc}})\right).$$

Not surprisingly, for the scale estimator Thall suggests the function

$$\lambda_{\mathrm{scale}}(\theta_{\mathrm{scale}}) = E\left(\psi_{\mathrm{scale}}\left(\frac{X}{\theta_{\mathrm{scale}}}\right)\right), \qquad \theta_{\mathrm{scale}} > 0.$$

Now let $\psi_{\mathrm{scale}} = \psi$ with ψ defined as in (3.4), then $|\psi(x)| \le b < \infty$ for all $x \in (0, \infty)$. That means $\lambda_{\mathrm{scale}}(\theta)$ certainly exists and is finite for all positive $\theta > 0$.

Due to the linearity of the integral and the continuity of ψ, $\lambda_{\mathrm{scale}}(\theta)$ is continuous. Since $\psi(x/\theta)$ is decreasing in θ, the same holds for $\lambda_{\mathrm{scale}}(\theta)$. Furthermore

$$\lim_{\theta \to 0} \lambda_{\mathrm{scale}}(\theta) = \lim_{\theta \to 0} \int\limits_0^\infty \max\left\{-b, \min\left\{\frac{s}{\theta} - (\alpha + a), b\right\}\right\} dF_\theta(s)$$

$$= \lim_{\theta \to 0} \left[- \int\limits_0^{(\alpha + a - b)\theta} b\, dF_\theta(s) + \int\limits_{(\alpha + a - b)\theta}^{(\alpha + a + b)\theta} \left(\frac{s}{\theta} - (\alpha + a)\right) dF_\theta(s) + \int\limits_{(\alpha + a + b)\theta}^\infty b F_\theta(s) \right]$$

$$= b > 0$$

and

$$\lim_{\theta \to \infty} \lambda_{\text{scale}}(\theta) = \lim_{\theta \to \infty} \int\limits_0^\infty \max\left\{-b, \min\left\{\frac{s}{\theta} - (\alpha + a), b\right\}\right\} dF_\theta(s)$$

$$= \lim_{\theta \to \infty} \left[- \int\limits_0^{(\alpha + a - b)\theta} b dF_\theta(s) + \int\limits_{(\alpha + a - b)\theta}^{(\alpha + a + b)\theta} \left(\frac{s}{\theta} - (\alpha + a)\right) dF_\theta(s) + \int\limits_{(\alpha + a + b)\theta}^\infty b F_\theta(s) \right]$$

$$= -b < 0.$$

By the Intermediate Value Theorem we conclude that there exists a $\theta_{0,i}$ such that $\lambda_{\text{scale}}(\theta_{0,i}) = 0$. Since ψ is monotone with range $[-b, b]$ and strictly monotone on $[\alpha + a - b, \alpha + a + b]$ we get $\lambda_{\text{scale}}(\theta) > 0$ for $\theta < \theta_{0,i}$ and $\lambda_{\text{scale}}(\theta) < 0$ for $\theta > \theta_{0,i}$ together with the linearity of the integral. Hence $T_{n,i}$ defined by

$$\sum_{j=1}^n \psi\left(\frac{X_{ij}}{T_{n,i}}\right) = 0$$

converges a.s. and in probability to $\theta_{0,i}$ according to Thall, Theorem 1.

It is further noticable that $\theta_{0,i}$ is unique due to Section 2.2.2.

So far we have shown

$$T_{n,i} \xrightarrow{P} \theta_{0,i} = T(F_{\theta_i}).$$

On the other hand, we know from Lemma 2.10 that T is Fisher-consistent yielding

$$T_{n,i} \xrightarrow{P} \theta_{0,i} = T(F_{\theta_i}) = \theta_i,$$

i.e. $T_{n,i}$ is consistent.

Even though the estimator $T_{n,i}$ given in (3.6) is of a more explicit form than the estimator $T_{n,i}$ in (3.4), we still would not be able to compute it directly. The reason is that the indices l_1 and l_2 depend on $T_{n,i}$, too, cf. (3.5). This means there is hardly no possibility to calculate $T_{n,i}$ in (3.6).

Now, instead of $T_{n,i}$ consider $\tilde{T}_{n,i}$ with

$$\tilde{T}_{n,i} = \frac{1}{(\alpha + a)d - (\alpha + a - b)\frac{\tilde{l}_1}{n} - (\alpha + a + b)\frac{\tilde{l}_2}{n}} \cdot \frac{1}{n} \sum_{j=\tilde{l}_1}^{n - \tilde{l}_2} X_{i(j)} \tag{3.9}$$

where $\tilde{l}_1, \tilde{l}_2 \in \mathbb{N}$ and $d \in \mathbb{R}$ are such that

$$\frac{X_{i(\tilde{l}_1)}}{\theta_i} \leq \alpha + a - b, \qquad \frac{X_{i(\tilde{l}_1 + 1)}}{\theta_i} > \alpha + a - b,$$

$$\frac{X_{i(n - \tilde{l}_2)}}{\theta_i} \leq \alpha + a + b, \qquad \frac{X_{i(n - \tilde{l}_2 + 1)}}{\theta_i} > \alpha + a + b,$$

and

$$(\alpha+a)d = \int_{\alpha+a-b}^{\alpha+a+b} s\,dF_1(s) + (\alpha+a-b)F_1(\alpha+a-b) + (\alpha+a+b)(1-F_1(\alpha+a+b)). \quad (3.10)$$

To closely analyse $\tilde{T}_{n,i}$ first note that the empirical distribution function $F_{n;\theta_i}$ evaluated at $(\alpha+a-b)\theta_i$ and $(\alpha+a+b)\theta_i$ for a sample $x_{i1}, \ldots, x_{in}$ is

$$F_{n;\theta_i}((\alpha+a-b)\theta_i; x_{i1}, \ldots, x_{in}) = \frac{1}{n}\sum_{j=1}^{n} \mathbb{1}_{(0,(\alpha+a-b]\theta_i]}(x_{ij})$$

$$F_{n;\theta_i}((\alpha+a+b)\theta_i; x_{i1}, \ldots, x_{in}) = \frac{1}{n}\sum_{j=1}^{n} \mathbb{1}_{(0,(\alpha+a+b]\theta_i]}(x_{ij}).$$

Since θ_i is a scale parameter, we also have

$$F_{n;\theta_i}((\alpha+a-b)\theta_i; x_{i1}, \ldots, x_{in}) = F_{n;1}(\alpha+a-b; x_{i1}, \ldots, x_{in})$$
$$F_{n;\theta_i}((\alpha+a+b)\theta_i; x_{i1}, \ldots, x_{in}) = F_{n;1}(\alpha+a+b; x_{i1}, \ldots, x_{in})$$

and by the Theorem of Glivenko-Cantelli (cf. [Müller 1991]) it follows

$$\frac{\tilde{l}_1}{n} = F_{n;1}(\alpha+a-b; X_1, \ldots, X_n) \xrightarrow{n\to\infty} F_1(\alpha+a-b)$$

$$\frac{n-\tilde{l}_2}{n} = F_{n;1}(\alpha+a+b; X_1, \ldots, X_n) \xrightarrow{n\to\infty} F_1(\alpha+a+b) \qquad \text{a.s.}$$

Next we examine the behaviour of $\frac{1}{n}\sum_{j=\tilde{l}_1}^{n-\tilde{l}_2} X_{i(j)}$.

Let $0 < l < u$ and X be a random variable with distribution function F. Then the random variable $Y = X \cdot \mathbb{1}_{[l,u]}(X)$ has distribution function

$$\begin{aligned}
G(y) &= P(X \cdot \mathbb{1}_{[l,u]}(X) \le y)\\
&= [P(X < l) + P(X > u)] \cdot \mathbb{1}_{[0,l)}(y) + [P(l \le X \le y) + P(X > u)] \cdot \mathbb{1}_{[l,u]}(y)\\
&\quad + \mathbb{1}_{(u,\infty)}(y)\\
&= 1 + F(y) \cdot \mathbb{1}_{[l,u]}(y) + F(l) \cdot \mathbb{1}_{[0,l)}(y) - F(l) \cdot \mathbb{1}_{[l,u]}(y) - F(u) \cdot \mathbb{1}_{[0,u]}(y).
\end{aligned}$$

Obviously, $dG(y) = \mathbb{1}_{[l,u]}(y)dF(y)$ since $F(l)$ and $F(u)$ are constants with respect to y. From the Weak Law of Large Numbers, we know

$$\frac{1}{n}\sum_{j=1}^{n} Y_{ij} \xrightarrow{P} \int_{0}^{\infty} y\,dG(y).$$

Now denote by $l(\tilde{l}_1)$, $u(\tilde{l}_2) \in \mathbb{R}$ real numbers such that

$$X_{i(\tilde{l}_1)} \le l(\tilde{l}_1) < X_{i(\tilde{l}_1+1)} \qquad \text{and} \qquad X_{i(n-\tilde{l}_2)} \le u(\tilde{l}_2) < X_{i(n-\tilde{l}_2+1)}$$

then

$$\frac{1}{n}\sum_{j=\tilde{l}_1}^{n-\tilde{l}_2} X_{i(j)} = \frac{1}{n}\sum_{j=1}^{n} X_{ij} \cdot \mathbb{1}_{[l(\tilde{l}_1),u(\tilde{l}_2)]}(X_{ij})$$

and choosing $l(\tilde{l}_1) = (\alpha + a - b)\theta_i$, $u(\tilde{l}_2) = (\alpha + a + b)\theta_i$ we get

$$\frac{1}{n}\sum_{j=1}^{n} X_{ij} \cdot \mathbb{1}_{[l(\tilde{l}_1),u(\tilde{l}_2)]}(X_{ij}) = \frac{1}{n}\sum_{j=1}^{n} X_{ij} \cdot \mathbb{1}_{[(\alpha+a-b)\theta,(\alpha+a+b)\theta_i]}(X_{ij}) = \frac{1}{n}\sum_{j=1}^{n} Y_{ij}$$

$$\xrightarrow{P} \int_0^\infty y dG(y) = \int_{l(\tilde{l}_1)}^{u(\tilde{l}_2)} x dF_\theta(x) = \int_{(\alpha+a-b)\theta_i}^{(\alpha+a+b)\theta_i} x dF_\theta(x) \overset{s\theta_i=x}{=} \theta_i \int_{\alpha+a-b}^{\alpha+a+b} s dF_1(s).$$

Thus

$$\tilde{T}_{n,i} \xrightarrow{P} \tilde{T}(F_{\theta_i})$$

with

$$\tilde{T}(F_1) = \frac{\int\limits_{\alpha+a-b}^{\alpha+a+b} s\,dF_1(s)}{(\alpha+a)d - (\alpha+a-b)F_1(\alpha+a-b) - (\alpha+a+b)(1 - F_1(\alpha+a+b))}.$$

Because of (3.10)

$$(\alpha+a)d = (\alpha+a-b)F_1(\alpha+a-b) + (\alpha+a+b)(1 - F_1(\alpha+a+b))$$
$$+ \int\limits_{\alpha+a-b}^{\alpha+a+b} s\,dF_1(s)$$

we have

$$(\alpha+a)d - (\alpha+a-b)F_1(\alpha+a-b) - (\alpha+a+b)(1 - F_1(\alpha+a+b)) = \int\limits_{\alpha+a-b}^{\alpha+a+b} s\,dF_1(s)$$

and $\tilde{T}(F_1) = 1$ as well as

$$\tilde{T}(F_{\theta_i}) = \theta_i \quad \tilde{T}(F_1) = \theta_i,$$

that is $\tilde{T}$ is Fisher-consistent. The scale equivariance of $\tilde{T}$ follows from Lemma 2.6.

Note that $\tilde{T}_{n,i}$ is defined by

$$\sum_{j=1}^{n} \tilde{\psi}\left(\frac{X_{ij}}{\tilde{T}_{n,i}}\right) = 0, \tag{3.11}$$

where

$$\tilde{\psi}(x) = \frac{1}{d} \begin{cases} \alpha + a - b - (\alpha + a)d, & x < \alpha + a - b \\ x - (\alpha + a)d, & \alpha + a - b \leq x \leq \alpha + a + b \\ \alpha + a + b - (\alpha + a)d, & x > \alpha + a + b \end{cases}$$

$$= \tfrac{1}{d} \left(\psi(x) + (\alpha + a) \right) - (\alpha + a) \tag{3.12}$$

Although we do have a Fisher-consistent M-estimator $\tilde{T}_{n,i}$, it is not optimal in the sense of Lemma 2.10 because of the definition of $\tilde{\psi}$.
Indeed, it seems questionable why $\tilde{T}_{n,i}$ is of better use than $T_{n,i}$. The indices $\tilde{l}_1$, $\tilde{l}_2$ in (3.9) do depend on θ_i, the parameter to be estimated. But recalling that $X/\theta_i \sim F_1$ for $X \sim F_{\theta_i}$, we can look at $\alpha + a - b$ and $\alpha + a + b$ as quantiles of the distribution function F_1. But then determing $\tilde{l}_1$ and $\tilde{l}_2$ is not anymore a matter of knowing θ_i or $\tilde{T}_{n,i}$ beforehand.

Kimber [Kimber 1983] presents the approach of defining $\tilde{T}_{n,i}$ by fixing some quantiles. Again the resulting generating function ψ_{quan} is not of the form (2.28) but has a similar shape.

For the remaining part of this chapter we omit the index i from $T_{n,i}$ because the assingment of the risk class i to its estimator $T_{n,i}$ is not necessary for the investigation to come up.

3.2.2   The Influence Functions of T and $\tilde{T}$

As it has been said in Section 2.3, one of the main tools in quantitative robustness is the influence function $IF(x; F, T)$, cf. (2.15).
We showed in Section 2.3.1 that for M-estimators, defined by a Gâteaux-differentiable functional, the influence function is directly proportional to the function ψ. Now we will examine the existence of the influence functions for both T and $\tilde{T}$.

As we pointed out for scale M-estimators T_n, the functional T is sufficiently described by (2.10), meaning we can evaluate (2.10) for the Gamma distribution $\Gamma(\alpha, 1)$.
Our M-estimators are defined by non-differentiable functions ψ (3.4) and $\tilde{\psi}$ (3.12). Nevertheless their influence functions exist, even though it is not easy to see.

Lemma 3.5
Let ψ be given as in (3.4). The functional T solves

$$\int_0^\infty \psi\left(\frac{s}{T(F_1)}\right) dF_1(s) = 0$$

with F_1 belonging to the class of Gamma distributions $\Gamma(\alpha, 1)$. Then

$$\lim_{t \to 0+} \frac{T((1-t)F_1 + t\Delta_x) - T(F_1)}{t} \tag{3.13}$$

exists and

$$IF(x; F_1, T) = \frac{\psi(x)}{\int\limits_{\alpha + a - b}^{\alpha + a + b} s \, dF_1(s)}. \tag{3.14}$$

Proof: To keep things readable, we denote $H_t = (1 - t)F_1 + t\Delta_x$ and thus

$$\int\limits_0^\infty \psi\left(\frac{s}{T(H_0)}\right) dH_0(s) = 0$$

by assumption.

Now, we consider the following expression

$$\frac{\int\limits_0^\infty \psi\left(\frac{s}{T(H_t)}\right) dH_t(s) - \int\limits_0^\infty \psi\left(\frac{s}{T(H_0)}\right) dH_0(s)}{t}$$

Appendix A.4 shows that from this we can derive

$$\lim_{t \to 0+} \frac{T((1 - t)F_1 + t\Delta_x) - T(F_1)}{t} = \frac{1}{\int\limits_{(\alpha + a - b)T(H_0)}^{(\alpha + a + b)T(H_0)} s \, dF_1(s)} \cdot \psi\left(\frac{x}{T(H_0)}\right) \cdot T^2(H_0).$$

Recalling that $T(H_0) = T(F_1) = 1$ because of the Fisher-consistency of T, the assumption follows immediately.   □

Since $\tilde{\psi}$ is a linear transformation of ψ, cf. (3.12), we can conclude the existence of the influence function of $\tilde{T}$ as well

$$IF(x; F_1, \tilde{T}) = \frac{\tilde{\psi}(x)}{\frac{1}{d} \int\limits_{\alpha + a - b}^{\alpha + a + b} s \, dF_1(s)}$$

$$= \frac{1}{\int\limits_{\alpha + a - b}^{\alpha + a + b} s \, dF_1(s)} \begin{cases} \alpha + a - b - (1 + a)d & x < \alpha + a - b \\ x - (1 + a)d & \alpha + a - b \le x \le \alpha + a + b \\ \alpha + a + b - (1 + a)d & x > \alpha + a + b. \end{cases}$$

$$\tag{3.15}$$

Remark 3.6

Note that Lemma 3.5 does not say, T and $\tilde{T}$ are Gâteaux-differentiable. Indeed, because H_t is a convex combination of F_1 and Δ_x, we actually consider only the limit from the right. Thus, ψ, $\tilde{\psi}$ being non-differentiable is not crucial anymore, since they are differentiable from the right.

In the last section of Chapter 3 we look for optimal values of a for both ψ and $\tilde{\psi}$.

3.3   The Optimal Choice for a

We will present and analyse two different approaches for determining an appropriate parameter a. Section 3.3.1 deduces a for ψ from condition (2.31) of Lemma 2.10. The second approach in Section 3.3.2 deals with the search for an optimal a for $\tilde{\psi}$. It follows the concept of minimising the M-estimator's asymptotic variance.

From now on – to stress the estimators' dependence on both a and b – we use the notation $T_n(a,b)$ and $\tilde{T}_n(a,b)$, respectively. Besides we restrict ourselves to the Exponential case, that is $\alpha = 1$. This restriction keeps analytical arguments capable.

3.3.1   Deriving a According to Lemma 2.10

We already pointed out, when stating Lemma 2.10, that (cf. (2.31))

$$\int_{\mathbb{R}} \psi(s;\theta)dF_\theta(s) = 0$$

yields a Fisher-consistent M-estimator, if the solution of

$$\int_{\mathbb{R}} \psi(s;T(F_\theta))dF_\theta(s) = 0$$

is unique.

For ψ chosen according to (3.4) in our case

$$\int_0^\infty \psi\left(\frac{s}{T(F_\theta;a,b)}\right)dF_\theta(s) = 0$$

has a unique zero, cf. p. 49. The idea we follow in this section is to find an $a \in \mathbb{R}$ for given $b > 0$ such that (2.31) in Lemma 2.10 is true. Remember that, because T_n is a scale M-estimator, (2.31) breaks down to

$$\int_0^\infty \psi(s)dF_1(s) = 0,$$

because $T(F_1) = 1$.

Before determining the optimal a with respect to the above condition, we gather some more information on the range of a.

Lemma 3.7
Let ψ be given as in Lemma 2.10 and let F_1 be the Exponential distribution $Exp(1)$. If the Fisher-consistent functional T is given by

$$\int_0^\infty \psi\left(\frac{s}{T(F_1;a,b)}\right)dF_1(s) = 0,$$

then $a \geq -1$.

Proof: Assume $1 + a < 0$, then certainly $1 + a - b < 0$ and

$$\int_0^\infty \psi\left(\frac{s}{T(F_1; a, b)}\right) dF_1(s) = \int_0^\infty \psi(s) dF_1(s)$$

$$= \int_0^{1+a+b} (s - (1+a))\, dF_1(s) + \int_{1+a+b}^\infty b\, dF_1(s)$$

$$= \int_0^{1+a+b} s\, dF_1(s) - (1+a)F_1(1+a+b) + b(1 - F_1(1+a+b))$$

$$> 0$$

because $1 + a < 0$. This is a contradiction. □

Now rewrite

$$0 = \int_0^\infty \psi(s) dF_1(s) = \int_0^{1+a-b} (-b)\, dF_1(s) + \int_{1+a-b}^{1+a+b} (s - (1+a))\, dF_1(s) + \int_{1+a+b}^\infty b\, dF_1(s).$$

and first consider the case $1 + a - b \leq 0$, that is $-1 < a \leq b - 1$. Then

$$\int_0^\infty \psi(s) dF_1(s) = \int_0^{1+a+b} (s - (1+a)) dF_1(s) + \int_{1+a+b}^\infty b\, dF_1(s)$$

$$= \int_0^{1+a+b} s e^{-s} ds - (1+a)(1 - e^{-(1+a+b)}) + b e^{-(1+a+b)}$$

$$= -a - e^{-(1+a+b)}$$

Thus

$$\int_0^\infty \psi(s) dF_1(s) = 0 \qquad \Longleftrightarrow \qquad e^{-(1+b)} = -a e^a. \tag{3.16}$$

We proceed by analysing the function $h_1(a) = -a e^a$ on $[-1, \infty)$, where the domain of h_1 is chosen according to the above lemma.

A simple calculation shows that h_1 has a unique extremal point – a global maximum – at $a_0 = -1$ with $h_1(a_0) = e^{-1}$. For $a > -1$, $h_1(a)$ is decreasing in a and $h_1(0) = 0$. Furthermore, because $b > 0$ we have $h_1(a_0) = e^{-1} > e^{-(1+b)} > 0 = h_1(0)$. Due to the continuity of h_1 we can conclude that there exists a unique a^* such that $e^{-(1+b)} = -a^* e^{a^*}$, we even know $a^* \in (-1, 0)$ for all $b > 0$.

To stress the connection to Fisher-consistency, we will denote the solution of (3.16) by a_{Fish}.

Now assume $1 + a - b > 0$, i.e. $a > b - 1$. Then

$$
\int\limits_0^\infty \psi(s)dF_1(s) = \int\limits_0^{1+a-b} (-b)e^{-s}ds + \int\limits_{1+a-b}^{1+a+b} (s - (1+a))e^{-s}ds + \int\limits_{1+a+b}^\infty be^{-s}ds
$$

$$
= -b - e^{-a}\left(e^{-(1+b)} - e^{-(1-b)}\right)
$$

that is

$$
\int\limits_0^\infty \psi(s)dF_1(s) = 0 \qquad \Longleftrightarrow \qquad be^a = e^{-(1-b)} - e^{-(1+b)}. \tag{3.17}
$$

In this case, we have a closer look at $h_2(a) = be^a$ on $(b - 1, \infty)$. Note that h_2 is strictly increasing on its domain.

Appendix A.5 shows that $h_2(b - 1) = be^{b-1} > e^{-(1-b)} - e^{-(1+b)}$ for $b > 1$. That means, if $b > 1$ there is no a in the domain of h_2, such that equation (3.17) will come true.

We are able to give an asymptotic value of parameter a_{Fish}.

Lemma 3.8
Let ψ be given by Lemma 2.10, $F_1 = Exp(1)$ and $T_n(a_{Fish}, b)$ be the Fisher-consistent scale M-estimator from (2.10), then

$$
\lim_{b\to\infty} \int\limits_0^\infty \psi\left(\frac{s}{T(F_1; a_{Fish}, b)}\right) dF_1(s)\Big|_{a_{Fish}=0} = 0.
$$

Proof: From (3.16) we conclude

$$
\lim_{b\to\infty} \int\limits_0^\infty \psi\left(\frac{s}{T(F_1; a, b)}\right) dF_1(s) = \lim_{b\to\infty} \left(-a - e^{-(1+a+b)}\right) = -a.
$$

Therefore

$$
\lim_{b\to\infty} \int\limits_0^\infty \psi\left(\frac{s}{T(F_1; a_{\text{Fish}}, b)}\right) dF_1(s) = 0 \qquad \Longleftrightarrow \qquad a_{\text{Fish}} = 0.
$$

$\square$

In Chapter 5 we will examine the behaviour of the M-estimator with both the exact value of parameter a_{Fish} given by (3.16) and the asymptotic value $a_{\text{asym}} = 0$.

Figure 3.1: Asymptotic Values for a_{Fish}

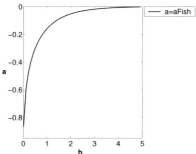

3.3.2   The Minimum Variance Approach

In Lemma 2.10 the function ψ was chosen such that the M-estimator $T_n(a,b)$ determined through

$$\int_{\mathbb{R}} \psi(s; T(F_1; a, b)) \, dF_1(s) = 0$$

is optimal in the sense that its asymptotic variance, cf. Section 2.3.3,

$$\text{Var}_{\text{asym}}(T_n(a,b)) = E(IF^2(X; F_1, T(a,b)))$$

is minimal among all Fisher-consistent M-estimators with bounded gross error.

Although the Fisher-consistent M-estimator $\tilde{T}_n(a,b)$ in (3.9) does not match the optimal form of Lemma 2.10, we take up the idea of minimising its asymptotic variance.

We will show that there is a unique $a_{\text{MV}} \in (-(b+1), \infty)$ for $\tilde{\psi}$ given in (3.12) such that for $b > 0$

$$a_{\text{MV}} = \underset{a \in \mathbb{R}}{\text{argmin}} \, E\left(IF^2(X; F_1, \tilde{T}(F_1; a, b)) \right).$$

Remark 3.0
So far we have not shown yet that

$$Var_{asym}(\tilde{T}_n) = E\left(IF^?(X; F_1, \tilde{T}(a, b)) \right)$$

as in (2.26). This will be done in Section 4.1.

For convenience we state $\tilde{\psi}$ here again (cf. (3.12))

$$\tilde{\psi}(x) = \frac{1}{d} \cdot \begin{cases} \alpha + a - b - (\alpha + a)d, & x < \alpha + a - b \\ x - (\alpha + a)d, & \alpha + a - b \le x \le \alpha + a + b \\ \alpha + a + b - (\alpha + a)d, & x > \alpha + a + b \end{cases}$$

and give the follwing results which hold for $\Gamma(\alpha, 1)$ in general. Detailed calculations can be found in Appendix A.6 and A.7

$$E(\psi(X)) = (\alpha + a)d - (\alpha + a)$$

$$E(\tilde{\psi}(X)) = E\left(\tfrac{1}{d}(\psi(X) + (\alpha + a))\right) - (\alpha + a) = 0$$

$$E\left(IF^2(X; F_1, \tilde{T}(a,b))\right) = \frac{Var(\psi(X))}{\left(\int\limits_{\alpha+a-b}^{\alpha+a+b} s\, dF_1(s)\right)^2}.$$

Now we return to the case $\alpha = 1$.

Remark 3.10
Note that because $F_1 = Exp(1)$ has support $(0, \infty)$ it is reasonable to ask for $1 + a + b > 0$ in

$$\int\limits_0^\infty \tilde{\psi}\left(\frac{s}{T(F_1; a, b)}\right) dF_1(s) = 0.$$

Thus we will look for an optimal $a \in (-(b+1), \infty)$.

For the theorem to come up, it is helpful to keep in mind that if $1 + a - b \leq 0$ we have

$$F_1(1 + a - b) = 0,$$

$$(\alpha + a)d(1 + a)d = (1 + a + b)(1 - F_1(1 + a + b)) + \int\limits_0^{1+a+b} s\, dF_1(s) \qquad (3.18)$$

and thus

$$E\left(IF^2(X; F_1, \tilde{T}(a,b))\right) = \frac{E\left(\tilde{\psi}^2(X)\right)}{\frac{1}{d}\left(\int\limits_0^{1+a+b} s\, dF_1(s)\right)^2}$$

$$= \frac{1}{\left(\int\limits_0^{1+a+b} s\, dF_1(s)\right)^2} \cdot \left[(1 + a + b)^2(1 - F_1(1 + a + b)) + \int\limits_0^{1+a+b} s^2\, dF_1(s)\right] - (1 + a)^2 d^2. \qquad (3.19)$$

If $1 + a - b > 0$ then $F_1(1 + a - b) > 0$ and

$$(1 + a)d = (1 + a - b)F_1(1 + a - b) + (1 + a + b)(1 - F_1(1 + a + b)) + \int\limits_{1+a-b}^{1+a+b} s\, dF_1(s) \qquad (3.20)$$

as well as

$$E\left(IF^2(X;F_1,\tilde{T}(a,b))\right) = \frac{1}{\left(\int\limits_{1+a-b}^{1+a+b} s\, dF_1(s)\right)^2} \cdot \left[(1+a-b)^2 F_1(1+a-b)\right.$$

$$+ \quad (1+a+b)^2(1 - F_1(1+a+b)) \tag{3.21}$$

$$\left. + \int\limits_{1+a-b}^{1+a+b} s^2\, dF_1(s) - (1+a)^2 d^2\right]. \tag{3.22}$$

Now we are able to prove

Theorem 3.11
Let $\tilde{T}$ be given by (2.10) and (3.12), that is

$$\tilde{\psi}(x) = \frac{1}{d} \cdot \begin{cases} 1+a-b-(1+a)d & x < 1+a-b \\ x-(1+a)d & 1+a-b \leq x \leq 1+a+b \\ 1+a+b-(1+a)d & x > 1+a+b \end{cases}$$

$$\int\limits_0^\infty \tilde{\psi}\left(\frac{s}{\tilde{T}(F_1;a,b)}\right) dF_1(s) = 0.$$

Then for all $b > 0$ there exists a unique $a_{MV} \in (-(b+1),\infty)$ such that

$$a_{MV} = \underset{a\in(-(b+1),\infty)}{\text{argmin}}\ \text{Var}\left(IF(X;F_1,\tilde{T}(a,b))\right).$$

Proof: Because of (2.23)

$$\text{Var}\left(IF(X;F_1,\tilde{T}(a,b))\right) = E\left(IF^2(X;F_1,\tilde{T}(a,b))\right).$$

The proof is divided into three steps.
At first we show that $E(IF^2(X;F_1,\tilde{T}(a,b)))$ is strictly monotone decreasing on the interval $(-(b+1),b-1]$ for all $b > 0$. Next, we prove that there exists a minimum of $E(IF^2(X;F_1,\tilde{T}(a,b)))$ in $(b-1,\infty)$ and finish the proof showing the uniqueness of the minimum.

We already discussed in Remark 3.10 that because the support of $Exp(1)$ is $(0,\infty)$, the range for a is $(-(b+1),\infty)$.

For $1+a-b \leq 0$, that is $-1 \leq a \leq b-1$, we have

$$F_1(1+a-b) = 0, \qquad F_1(1+a+b) = 1 - e^{-(1+a+b)}$$

and according to (3.18)

$$(1+a)d = 1 - e^{-(1+a+b)}.$$

Calculating the asymptotic variance $\mathrm{Var}(IF(X; F_1, \tilde{T}(a,b))) = E(IF^2(X; F_1, \tilde{T}(a,b)))$, we recall that

$$\int_0^{1+a+b} s\, dF_1(s) = \int_0^{1+a+b} s e^{-s}\, ds = 1 - (2+a+b)e^{-(1+a+b)}$$

$$\int_0^{1+a+b} s^2\, dF_1(s) = \int_0^{1+a+b} s^2 e^{-s}\, ds = 2 - ((2+a+b)^2 + 1)e^{-(1+a+b)}$$

and applying (3.19) yields

$$E\left(IF^2(X; F_1, \tilde{T}(a,b))\right)$$

$$= \frac{1}{\left(1 - (2+a+b)e^{-(1+a+b)}\right)^2}$$

$$\cdot \left[(1+a+b)^2 e^{-(1+a+b)} + 2 - ((2+a+b)^2 + 1)e^{-(1+a+b)} - \left(1 - e^{-(1+a+b)}\right)^2 \right]$$

$$= \frac{1 - 2(1+a+b)e^{-(1+a+b)} - e^{-2(1+a+b)}}{\left(1 - (2+a+b)e^{-(1+a+b)}\right)^2}$$

$$= \frac{e^{2(1+a+b)} - 2(1+a+b)e^{1+a+b} - 1}{\left(e^{1+a+b} - (2+a+b)\right)^2}. \tag{3.23}$$

Since we want to show that $E(IF^2(X; F_1, \tilde{T}(a,b)))$ is strictly monotone decreasing, we look at the first derivative of the variance with respect to a.

$$\frac{\partial}{\partial a} E\left(IF^2(X; F_1, \tilde{T}(a,b))\right)$$

$$= \frac{2}{(e^{1+a+b} - (2+a+b))^3}$$

$$\cdot \left[\left(e^{1+a+b} - (2+a+b)\right)\left(e^{2(1+a+b)} - e^{1+a+b} - (1+a+b)e^{1+a+b}\right) \right.$$

$$\left. - \left(e^{1+a+b} - 1\right)\left(e^{2(1+a+b)} - 2(1+a+b)e^{1+a+b} - 1\right) \right]$$

$$= \frac{2}{(e^{1+a+b} - (2+a+b))^3}$$

$$\cdot \left[-(1+a+b)e^{2(1+a+b)} - (2+a+b)e^{2(1+a+b)} \right.$$

$$+ (2+a+b)(1+a+b)e^{1+a+b} + (2+a+b)e^{1+a+b}$$

$$\left. + 2(1+a+b)e^{2(1+a+b)} + e^{1+a+b} - 2(1+a+b)e^{1+a+b} - 1 \right]$$

$$= 2\frac{-e^{2(1+a+b)} + e^{1+a+b}\left(2 + (1+a+b)^2\right) - 1}{(e^{1+a+b} - (2+a+b))^3}. \tag{3.24}$$

The denominator of the above fraction is always positive. To show this, observe that $x \mapsto 1+x$ is the tangent of $x \mapsto e^x$ in $x = 0$. Further $x \mapsto e^x$ is strictly convex.

Now choose $x = 1 + a + b$.

Because the denominator is positive, the numerator has to be negative to conclude that $E(IF^2(X; F_1, \tilde{T}(a,b)))$ is strictly monotone decreasing. Thus, we check whether

$$-e^{2(1+a+b)} + e^{1+a+b}(2 + (1+a+b)^2) - 1 = -\left(e^{1+a+b} - 1\right)^2 + (1+a+b)^2 e^{1+a+b} < 0.$$

The statement is true if

$$(1 + a + b)^2 e^{1+a+b} < \underbrace{(e^{1+a+b} - 1)^2}_{>0}$$

$$\Longleftrightarrow \quad (1 + a + b)e^{1/2(1+a+b)} < e^{1+a+b} - 1$$

$$\Longleftrightarrow \quad \frac{1+a+b}{2} < \sinh\left(\frac{1+a+b}{2}\right).$$

Because $1 + a + b > 0$ the last statement is true, cf. Appendix 8.

This means

$$\frac{\partial}{\partial a} E\left(IF^2(X; F_1, \tilde{T}(a,b))\right) < 0$$

and we conclude that the asymptotic variance of the M-estimator $\tilde{T}_n$ is strictly decreasing for $a \in (-(b+1), b-1]$.

We continue the proof, showing that the asymptotic variance $E(IF^2(X; F_1, \tilde{T}(a,b)))$ has a minimum with respect to a in $(b-1, \infty)$.

Let $1 + a - b > 0$, that is $a > b - 1$. Then

$$F_1(1 + a - b) = 1 - e^{-(1+a-b)}, \qquad F_1(1 + a + b) = 1 - e^{-(1+a+b)}$$

and

$$\int_{1+a-b}^{1+a+b} s \, dF_1(s) = \int_{1+a-b}^{1+a+b} s e^{-s} \, ds = (2 + a - b)e^{-(1+a-b)} - (2 + a + b)e^{-(1+a+b)}$$

$$\int_{1+a-b}^{1+a+b} s^2 \, dF_1(s) = \int_{1+a-b}^{1+a+b} s^2 e^{-s} \, ds$$

$$= \left((2 + a - b)^2 + 1\right) e^{-(1+a-b)} - \left((2 + a + b)^2 + 1\right) e^{-(1+a+b)}.$$

Therefore the asymptotic variance of $\tilde{T}_n$ is given by

$$E\Big(IF^2(X; F_1, \tilde{T}(a,b))\Big)$$

$$= \frac{1}{((2+a-b)e^{-(1+a-b)} - (2+a+b)e^{-(1+a+b)})^2}$$

$$\cdot \Big[(1+a-b)^2(1 - e^{-(1+a-b)}) + (1+a+b)^2 e^{-(1+a+b)}$$

$$+ ((2+a-b)^2 + 1)e^{-(1+a-b)} - ((2+a+b)^2 + 1)e^{-(1+a+b)}$$

$$- \Big((1+a-b)(1 - e^{-(1+a-b)}) + (1+a+b)e^{-(1+a+b)}$$

$$+ (2+a-b)e^{-(1+a-b)} - (2+a+b)e^{-(1+a+b)}\Big)^2\Big]$$

$$= \frac{2e^{1+a+3b} - 2e^{1+a+b} - 2(1+a+b)e^{1+a+b} + 2(1+a-b)e^{1+a+b} - (e^{2b}-1)^2}{(e^{2b}(2+a-b) - (2+a+b))^2}$$

$$= \frac{2e^{1+a+b}(e^{2b} - (1+2b)) - (e^{2b}-1)^2}{(e^{2b}(2+a-b) - (2+a+b))^2}, \tag{3.25}$$

because of (3.20) together with (3.22).

To show the existence of an $a \in (b-1, \infty)$ minimising $E(IF^2(X; F_1, \tilde{T}(a,b)))$, we prove that

$$a \mapsto \frac{2e^{1+a+b}(e^{2b} - (1+2b)) - (e^{2b}-1)^2}{(e^{2b}(2+a-b) - (2+a+b))^2}, \qquad a > b-1$$

is a continuous function taking infinitely large values as $a \to \infty$. Together with our results from part one of this proof, we then draw the desired conclusions.

The above function is well defined as it can easily be seen. Note that the denominator is an open-upward parabola in a. And since

$$\frac{\partial}{\partial a}\left(e^{2b}(2+a-b) - (2+a+b)\right)^2\Big|_{a=b-1}$$

$$= 2\left(e^{2b}(2+a-b) - (2+a+b)\right)\left(e^{2b} - 1\right)\Big|_{a=b-1}$$

$$= 2\left(e^{2b} - (1+2b)\right)\left(e^{2b} - 1\right)$$

$$> 0$$

the quadratic function $a \mapsto \left(e^{2b}(2+a-b) - (2+a+b)\right)^2$ is strictly increasing on $(b-1, \infty)$. Hence the parabola's only zero $a_0 = -(e^{2b}(2-b) - (2+b))/(e^{2b}-1)$ lies in $(-\infty, b-1]$.

In addition

$$\lim_{a \to \infty} E\Big(IF^2(X; F_1, \tilde{T}(a,b))\Big) = \lim_{a \to \infty} \frac{2e^{1+a+b}(e^{2b} - (1+2b)) - (e^{2b}-1)^2}{(e^{2b}(2+a-b) - (2+a+b))^2} = \infty$$

since $e^{2b} - (1 + 2b) > 0$ due to convexity (cf. part one of this proof) and

$$E\left(IF^2(X; F_1, \tilde{T}(a, b))\right) \asymp \frac{e^a}{a^2} \xrightarrow{a \to \infty} \infty.$$

Furthermore $E(IF^2(X; F_1, \tilde{T}(a, b)))$ is continuous since

$$\lim_{a \to (b-1)-} E\left(IF^2(X; F_1, \tilde{T}(a, b))\right) = \frac{e^{2(1+a+b)} - 2(1 + a + b)e^{1+a+b} - 1}{(e^{1+a+b} - (2 + a + b))^2}\Bigg|_{a=b-1}$$

$$= \frac{e^{4b} - 4be^{2b} - 1}{(e^{2b} - (1 + 2b))^2},$$

cf. (3.23) and according to (3.25)

$$\lim_{a \to (b-1)+} E\left(IF^2(X; F_1, \tilde{T}(a, b))\right) = \frac{2e^{1+a+b}(e^{2b} - (1 + 2b)) - (e^{2b} - 1)^2}{(e^{2b}(2 + a - b) - (2 + a + b))^2}\Bigg|_{a=b-1}$$

$$= \frac{e^{4b} - 4be^{2b} - 1}{(e^{2b} - (1 + 2b))^2}.$$

Part one of this proof showed that $E(IF^2(X; F_1, \tilde{T}(a, b)))$ is decreasing on $(-(b+1), b-1]$. In addition, we just proved $E(IF^2(X; F_1, \tilde{T}(a, b)))$ to be continuous on $(-(b+1), \infty)$ and to converge to infinity for $a \to \infty$.
Putting everything together, we conclude that there exists $a_{\mathrm{MV}} \in (b-1, \infty)$ minimising $E(IF^2(X; F_1, \tilde{T}(a, b)))$.

We are left with proving the uniqueness of a_{MV}. For this to show, we again examine the behaviour of the first derivative of $E(IF^2(X; F_1, \tilde{T}(a, b)))$ with respect to $a \in (b-1, \infty)$. Recalling (3.25)

$$\frac{\partial}{\partial a} E\left(IF^2(X; F_1, \tilde{T}(a, b))\right)$$

$$= \frac{1}{(e^{2b}(2 + a - b) - (2 + a + b))^3}$$

$$\cdot \left[2e^{1+a+b}\left(e^{2b} - (1 + 2b)\right)\left(e^{2b}(2 + a - b) - (2 + a + b)\right)\right.$$

$$\left. - 2\left(e^{2b} - 1\right)\left(2e^{1+a+b}(e^{2b} - (1 + 2b)) - (e^{2b} - 1)^2\right)\right]$$

$$= \frac{2}{(e^{2b}(2 + a - b) - (2 + a + b))^3}$$

$$\cdot \left[\left(e^{2b} - 1\right)^3 + e^{1+a+b}\left(e^{2b} - (1 + 2b)\right)\left(e^{2b}(2 + a - b) - (2 + a + b) - 2(e^{2b} - 1)\right)\right]$$

$$\tag{3.26}$$

The idea behind the following proof of uniqueness is to show that the first derivative of $E(IF^2(X; F_1, \tilde{T}(a, b)))$ with respect to a is strictly convex on $(b-1, \infty)$. It ranges from negative to positive values. Then it follows immediately that there has to be a unique zero of $(\partial/\partial a)E(IF^2(X;F_1,\tilde{T}(a,b)))$ in $(b-1,\infty)$. In connection with the results from part one and two of this proof, we can conlcude that there is a unique a_{MV} minimising $E(IF^2(X;F_1,\tilde{T}(a,b)))$.

The denominator is greater than zero since $a \mapsto (e^{2b}(2+a-b)-(2+a+b))^3$ has the same zero a_0 as $a \mapsto (e^{2b}(2+a-b)-(2+a+b))^2$ from above, i.e. $a_0 < b-1$. Thus any possible change of sign in (3.26) is caused by the numerator

$$h(a,b) := \left(e^{2b}-1\right)^3 + e^{1+a+b}\left(e^{2b}-(1+2b)\right)\left(e^{2b}(2+a-b)-(2+a+b)-2(e^{2b}-1)\right).$$

We will conclude that the first derivative of $E(IF^2(X;F_1,\tilde{T}(a,b)))$ has a unique zero by showing that h is strictly convex in a with values ranging from negative to positive values.

The terms $\left(e^{2b}-1\right)^3$ and $\left(e^{2b}-(1+2b)\right)$ of $h(a,b)$ are constants with respect to a and can therefore be omitted in the following investigation. Obviously h is continuous in a.

Note that

$$\frac{\partial}{\partial a}\left[e^{1+a+b}\left(e^{2b}(2+a-b)-(2+a+b)-2(e^{2b}-1)\right)\right]$$
$$= e^{1+a+b}\left(e^{2b}(1+a-b)-(1+a+b)\right)$$

and

$$\frac{\partial^2}{\partial a^2}\left[e^{1+a+b}\left(e^{2b}(2+a-b)-(2+a+b)-2(e^{2b}-1)\right)\right]$$
$$= e^{1+a+b}\left(e^{2b}(2+a-b)-(2+a+b)\right).$$

The last expression is positive for $b > 0$ because

$$\left[e^{2b}(2+a-b)-(2+a+b)\right]_{b=0} = 0$$
$$\frac{\partial}{\partial b}\left[e^{2b}(2+a-b)-(2+a+b)\right] = e^{2b}(3+2a-2b)-1$$
$$= e^{2b}(1+2(1+a-b))-1 > 0$$

since $1+a-b > 0$ by assumption.
Now, this means $(\partial/\partial a)h(a,b) > 0$, i.e. $h(.,b)$ is strictly convex for $a > b-1$.

The next steps show that $h(.,b)$ ranges from negative to positive values as a goes from $b-1$ to ∞. At first, we notice

$$\lim_{a\to(b-1)+} h(a,b) = \left(e^{2b}-1\right)^3 + e^{2b}\left(e^{2b}-(1+2b)\right)\left(e^{2b}-(1+2b)-2(e^{2b}-1)\right)$$

and

$$\lim_{b\to 0} h(a,b) = 0.$$

Furthermore (cf. Appendix A.9)

$$\frac{\partial}{\partial b} \left[\lim_{a \to (b-1)+} h(a,b) \right] = \frac{d}{db} \left(- \left(e^{2b} - 1 \right)^2 + 4b^2 e^{2b} \right)$$
$$= 4e^{2b} \left(-e^{2b} + 1 + 2b + 2b^2 \right) < 0$$

because of the Taylor expansion of e^{2b}. That is, $\lim_{a \to (b-1)+}$ is decreasing in b. Therefore we can conlude $h(b-1,b) < 0$ for $b > 0$.

Now, define $\bar{a} = b \cdot \frac{e^{2b}+1}{e^{2b}-1} > b - 1$, then

$$h(\bar{a},b) = (e^{2b} - 1)^3 + e^{1+\bar{a}+b} \left(e^{2b} - (1+2b) \right) \left(e^{2b}(2 + \bar{a} - b) - (2 + \bar{a} + b) - 2(e^{2b} - 1) \right)$$
$$= \left(e^{2b} - 1 \right)^3 + e^{1+\bar{a}+b} \left(e^{2b} - 2b - 1 \right) \left(e^{2b}(\bar{a} - b) - (\bar{a} + b) \right)$$
$$= \left(e^{2b} - 1 \right)^3 > 0$$

and for any $a_1 > \bar{a}$ we get

$$e^{2b}(a_1 - b) - (a_1 + b) = a_1 \left(e^{2b} - 1 \right) - b \left(e^{2b} + 1 \right) > \bar{a} \left(e^{2b} - 1 \right) - b \left(e^{2b} + 1 \right)$$

and hence $h(a_1, b) > h(\bar{a}, b) > 0$.
So we have $\lim_{a \to (b-1)+} h(a,b) < 0$ and $h(a_1, b) > 0$ for $a_1 > \bar{a}$, $b > 0$.

Because of its continuity and strict convexity with respect to a, we know that $h(.,b)$ has a unique zero in $(b-1, a_1)$. As $a_1 > \bar{a}$ is chosen arbitrarily this implies there exists exactly one zero in $(b-1, \infty)$.

But this means $\frac{\partial}{\partial a} E(IF^2(X; F_1, \tilde{T}(a,b)))$ has a unique zero in $(b-1, \infty)$.
Therefore $E(IF^2(X; F_1, \tilde{T}(a,b)))$ has a unique minimum in $(b-1, \infty)$. □

Even though the optimal a_{MV} cannot be stated explicitly, we can give an asymptotic value.

Lemma 3.12
Let $\tilde{\psi}$ be given by (3.12) and let $\tilde{T}_n$ be the scale M-estimator with functional $\tilde{T}$ given by (2.10), then

$$\lim_{b \to \infty} \frac{\partial}{\partial a_{\mathrm{MV}}} E \left(IF^2(X; F_1, \tilde{T}(a_{\mathrm{MV}}, b)) \right) \bigg|_{a_{\mathrm{MV}}=b-1} = 0.$$

Figure 3.2: Asymptotic Values for a_{MV}

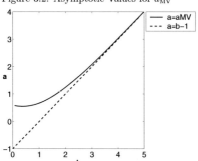

Proof:　First assume $1 + a_{\mathrm{MV}} \leq 0$. Applying (3.24) we get

$$\frac{\partial}{\partial a_{\mathrm{MV}}} E\left(IF^2(X; F_1, \tilde{T}(a_{\mathrm{MV}}, b))\right)\Big|_{a=b-1}$$

$$= 2\frac{-e^{2(1+a_{\mathrm{MV}}+b)} + e^{1+a_{\mathrm{MV}}+b}(2 + (1 + a_{\mathrm{MV}} + b)^2) - 1}{(e^{1+a_{\mathrm{MV}}+b} - (2 + a_{\mathrm{MV}} + b))^3}\Big|_{a_{\mathrm{MV}}=b-1}$$

$$= 2\frac{-e^{4b} + 2e^{2b}(1 + 2b^2) - 1}{(e^{2b} - (1 + 2b))^3}$$

$$= 2\frac{-\left(e^{2b} - 1\right)^2 + 4b^2 e^{2b}}{(e^{2b} - (1 + 2b))^3}.$$

If $1 + a_{\mathrm{MV}} - b > 0$ then according to (3.26)

$$\frac{\partial}{\partial a_{\mathrm{MV}}} E\left(IF^2(X; F_1, \tilde{T}(a_{\mathrm{MV}}, b))\right)\Big|_{a_{\mathrm{MV}}=b-1} = 2\frac{-\left(e^{2b} - 1\right)^2 + 4b^2 e^{2b}}{(e^{2b} - (1 + 2b))^3}.$$

Finally

$$\lim_{b \to \infty} 2\frac{-\left(e^{2b} - 1\right)^2 + 4b^2 e^{2b}}{(e^{2b} - (1 + 2b))^3} = \lim_{b \to \infty} 2\frac{e^{8b}\left(-\left(e^{-2b} - e^{-4b}\right)^2 + 4b^2 e^{-6b}\right)}{e^{8b}\left(1 - e^{-2b}(1 + 2b)\right)^3} = 0.$$

$\square$

In Chapter 5, we will examine whether there is a difference between the application of a_{MV} and $a_{\mathrm{asym}} = b - 1$.

But before doing the simulation, we have a closer look at some quantitative characteristics of T_n and $\tilde{T}_n$.

Chapter 4

Quantitative Characteristics of T_n and $\tilde{T}_n$

Section 2.3 told us about different characteristics that can be applied to judge an estimator's robustness in quantitative terms. In this chapter we will determine some of these quantities. Indeed, calculating the finite sample breakdown points in Section 4.3, we also discuss the problem of choosing reasonable values for b in ψ and $\tilde{\psi}$.

Because now the emphasis does not lie on the value of a and b anymore, we go back to our original notations $T_{n,i}$ and $\tilde{T}_{n,i}$ in Sections 4.1, 4.3 and T_n, $\tilde{T}_n$ in Section 4.2. However, we start with a closer inspection of the asymptotic behaviour of both $T_{n,i}$ and $\tilde{T}_{n,i}$. Note that from now on T_n, $\tilde{T}_n$ ($T_{n,i}$, $\tilde{T}_{n,i}$) are M-estimators of the parameter θ (θ_i) of the Exponential distribution $Exp(\theta)(Exp(\theta_i))$.

4.1 Asymptotic Normality of $T_{n,i}$ and $\tilde{T}_{n,i}$

In preparation of finding the function ψ yielding an estimator $T_{n,i}$ for risk parameter $\theta_i, i = 1, \ldots, N$ with minimal asymptotic variance σ^2_{asym}, compare (2.26) and Lemma 2.10, we already discussed the question of asymptotic normality. One of the essential requirements of the approach to gain asymptotic normality in Section 2.3.3 was, that for the distribution function F_θ the condition $\sqrt{n}||F_n - F_\theta|| = o(1)$ is true.

The line we are going to take now, dispenses this condition. Instead we are applying Theorem 2 from Thall [Thall 1979].

Theorem 4.1 ([Thall 1979], Theorem 2)
Let $\lambda : (0, \infty) \to [0, \infty)$,

$$\lambda(t) = \int\limits_0^\infty \psi\left(\tfrac{s}{t}\right) dF_{\theta_i}(s)$$

and suppose there exists a value $c \in (0, \infty)$ *such that*

(i) $\lambda(c) = 0$,

(ii) λ is differentiable at $t = c$ and $\lambda'(c) < 0$,

(iii) at $t = c$ the integral $\int_0^\infty \psi^2 \left(\frac{s}{t}\right) dF_{\theta_0}(s)$ is finite and continuous.

Let $T_{n,i}$ be the M-estimator of θ_i solving

$$\sum_{j=1}^n \psi \left(\frac{X_{ij}}{T_{n,i}}\right) = 0,$$

then $\sqrt{n}(T_n - c)$ converges in distribution to a normal random variable with mean 0 and variance

$$\sigma_{asym}^2 = \frac{1}{(\lambda'(c))^2} \int_0^\infty \psi^2 \left(\frac{s}{c}\right) dF_{\theta_0}(s).$$

We will apply the above theorem to the estimator $T_{n,i}$ and then show by a simple calculation that we can derive asymptotic normality for $\tilde{T}_{n,i}$ as well. Recalling the definition of $T_{n,i}$ in (3.4), that is

$$\sum_{j=1}^n \psi \left(\frac{X_{ij}}{T_{n,i}}\right) = 0, \qquad \psi(x) = \max\{-b, \min\{x - (1+a), b\}\},$$

and due to its consistency, cf. Section 3.2.1, we can conclude that condition (i) is fulfilled for $c = T(F_{\theta_i})$ at the true model distribution function F_{θ_i}.
To see that λ is differentiable at $t = c$ and $\lambda'(c) < 0$ note that

$$\lambda(t) = \int_0^\infty \psi \left(\frac{s}{t}\right) dF_{\theta_i}(s)$$

$$= - \int_0^{(1+a-b)t} b \, dF_{\theta_i}(s) + \int_{(1+a-b)t}^{(1+a+b)t} \left(\frac{s}{t} - (1+a)\right) dF_{\theta_i}(s) + \int_{(1+a+b)t}^\infty b \, dF_{\theta_i}(s)$$

$$= (1+a-b)F_{\theta_i}((1+a-b)t) + (1+a+b)(1 - F_{\theta_i}((1+a+b)t)) - (1+a)$$

$$+ \int_{(1+a-b)t}^{(1+a+b)t} \frac{s}{t} \, dF_{\theta_i}(s)$$

is obviously differentiable for all $t > 0$ since F_{θ_i} is a continuous distribution function. Furthermore

$$\lambda'(t) = (1+a-b)^2 f_{\theta_i}((1+a-b)t) - (1+a+b)^2 f_{\theta_i}((1+a+b)t)$$

$$- (1+a-b)^2 f_{\theta_i}((1+a-b)t) + (1+a+b)^2 f_{\theta_i}((1+a+b)t) - \int_{(1+a-b)t}^{(1+a+b)t} \frac{s}{t^2} \, dF_{\theta_i}(s)$$

$$= - \int\limits_{(1+a-b)t}^{(1+a+b)t} \frac{s}{t^2}\, dF_{\theta_i}(s) < 0$$

for all $t > 0$ since $(1 + a + b)t > 0$ according to our discussion in Section 3.3.2. We are left proving condition (iii). Observe that

$$\eta(t) := \int\limits_0^\infty \psi^2\left(\tfrac{s}{t}\right) dF_{\theta_i}(s)$$

$$= b^2 F_{\theta_i}\big((1 + a - b)t\big) + b^2\big(1 - F_{\theta_i}((1 + a + b)t)\big) + \int\limits_{(1+a-b)t}^{(1+a+b)t} \left(\frac{s}{t} - (1 + a)\right)^2 dF_{\theta_i}(s)$$

$$< \infty$$

since $s \mapsto (s/t - (1 + a))^2$ is a continuous function for any $t > 0$ and is integrated over an interval of finite length. Because F_{θ_i} is a continuous distribution function, $t \mapsto \eta(t)$ is also continuous on $(0, \infty)$. So by Thall's Theorem 2 we have shown asymptotic normality of $T_{n,i}$.

From Thall's Theorem 2 it also follows that

$$\sigma^2_{\text{asym}}(T_{n,i}) = \frac{\int\limits_0^\infty \psi^2\left(\tfrac{s}{c}\right) dF_{\theta_i}(s)}{\left(\frac{\partial}{\partial t} \int\limits_0^\infty \psi\left(\tfrac{s}{t}\right) dF_{\theta_i}\big|_{t=c}\right)^2}$$

with $c = T(F_{\theta_i})$. Now

$$\frac{\partial}{\partial t} \int\limits_0^\infty \psi\left(\frac{s}{t}\right) dF_{\theta_i}(s) = \frac{\partial}{\partial t}(1 + a - b)F_{\theta_i}\big((1 + a - b)t\big)$$

$$+ (1 + a + b)\big(1 - F_{\theta_i}((1 + a + b)t)\big) - (1 + a)$$

$$+ \int\limits_{(1+a-b)t}^{(1+a+b)t} \frac{s}{t}\, dF_{\theta_i}(s)$$

$$= - \int\limits_{(1+a-b)t}^{(1+a+b)t} \frac{s}{t^2}\, dF_{\theta_i}(s),$$

yielding

$$\sigma^2_{\text{asym}}(T_{n,i}) = \frac{\int\limits_0^\infty \psi^2\left(\frac{s}{T(F_{\theta_i})}\right) dF_{\theta_i}(s) \cdot T(F_{\theta_i})}{\left(\int\limits_{(1+a-b)T(F_{\theta_i})}^{(1+a+b)T(F_{\theta_i})} \frac{s}{T(F_{\theta_i})}\, dF_{\theta_i}(s)\right)^2} = E\left(IF^2(X; F_{\theta_i}, T)\right). \qquad (4.1)$$

Therefore the estimator $T_{n,i}$ in (3.4) is chosen to have minimum asymptotic variance, cf. Lemma 2.10.

To gain asymptotic normality of $\tilde{T}_{n,i}$ recall (3.12),

$$\tilde{\psi}(x) = \tfrac{1}{d}\left(\psi(x) + (1+a)\right) - (1+a).$$

Define

$$\tilde{\lambda}(t) := \int\limits_0^\infty \tilde{\psi}\left(\frac{s}{t}\right) dF_{\theta_i}(s),$$

then $\tilde{\lambda}(\tilde{T}(F_{\theta_i})) = 0$. And since

$$\tilde{\lambda}(t) = \int\limits_0^\infty \tilde{\psi}\left(\frac{s}{t}\right) dF_{\theta_i}(s) = \frac{1}{d}\int\limits_0^\infty \left(\frac{s}{t}\right) dF_{\theta_i}(s) + \frac{1}{d}(1+a) - (1+a)$$

$$= \tfrac{1}{d}(\lambda(t) + (1+a)) - (1+a)$$

we can transfer our results from $\lambda(t)$ to $\tilde{\lambda}(t)$ and the asymptotic normality of $\tilde{T}_{n,i}$ follows with

$$\sigma^2_{\text{assym}}(\tilde{T}_n) = \frac{1}{(\tilde{\lambda}(\tilde{T}(F_{\theta_i})))^2} \cdot \int\limits_0^\infty \tilde{\psi}^2 \left(\frac{s}{\tilde{T}(F_{\theta_i})}\right)^2 dF_{\theta_i}(s) = E\left(IF^2(X; F_{\theta_i}, \tilde{T})\right). \quad (4.2)$$

4.2   Gross Error

In this section we again omit the risk class index i.
Recall from Section 2.3, (2.16), the gross error of some estimator T_n of parameter θ is defined as

$$\gamma^*(T) = \sup_{x \in (0,\infty)} |IF(x; F_\theta, T)|.$$

Before we calculate the gross errors of T_n and $\tilde{T}_n$, we state and sketch the influence functions of both estimators, cf. (3.14) and (3.15).

$$IF(x; F_\theta, T) = \theta \cdot IF(x; F_1, T) \qquad \text{according to (2.21)}$$

$$IF(x; F_1, T) = \frac{\psi(x)}{\frac{1}{1+a+b}\int\limits_{1+a-b}^{} s\, dF_1(s)}$$

$$= \frac{1}{\frac{1}{1+a+b}\int\limits_{1+a-b}^{} s\, dF_1(s)} \begin{cases} -b, & x < 1+a-b \\ x - (1+a), & 1+a-b \le x \le 1+a+b \\ b, & x > 1+a+b. \end{cases}$$

Figure 4.1: Influence Function $IF(x; F_1, T(a, 1))$ for T, $b = 1$

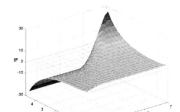

Figure 4.2: Influence Function $IF(x; F_1, T(0.5, b))$ for T, $a = 0.5$

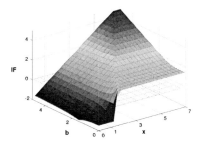

$$IF(x; F_\theta, \tilde{T}) - \theta \cdot IF(x; F_1, \tilde{T})$$

$$IF(x; F_1, \tilde{T}) = \frac{\tilde{\psi}(x)}{\frac{1}{d} \int\limits_{1+a-b}^{1+a+b} s \, dF_1(s)}$$

$$= \frac{1}{\int\limits_{1+a-b}^{1+a+b} s \, dF_1(s)} \begin{cases} 1 + a - b - (1+a)d, & x < 1 + a - b \\ x - (1+a)d, & 1 + a - b \leq x \leq 1 + a + b \\ 1 + a + b - (1+a)d, & x > 1 + a + b. \end{cases}$$

Figure 4.3: Influence Function $IF(x; F_1, \tilde{T}(a, 1))$ for $\tilde{T}$, $b = 1$

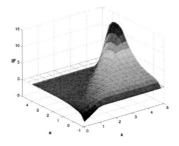

Figure 4.4: Influence Function $IF(x; F_1, \tilde{T}(0.5, b))$ for $\tilde{T}$, $a = 0.5$

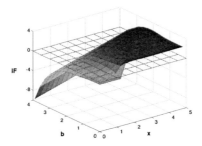

It follows that

$$\gamma^*(T(F_\theta)) = \sup_{x \in (0,\infty)} |\theta \cdot IF(x; F_1, T)| = \theta \cdot \frac{b}{\int\limits_{1+a-b}^{1+a+b} s\, dF_1(s)}$$

because $1 + a > 0$ and

$$\gamma^*(\tilde{T}(F_\theta)) = \sup_{x \in (0,\infty)} |\theta \cdot IF(x; F_1, \tilde{T})| = \theta \cdot \frac{|1 + a + b - (1 + a)d|}{\int\limits_{1+a-b}^{1+a+b} s\, dF_1(s)}.$$

Since

$$(1+a)d = (1+a-b)F_1(1+a-b) + (1+a+b)(1-F_1(1+a+b)) + \int\limits_{1+a-b}^{1+a+b} s\, dF_1(s)$$

$$= -[(1+a+b)F_1(1+a+b) - (1+a-b)F_1(1+a-b)] + (1+a+b)$$

$$+ \int\limits_{1+a-b}^{1+a+b} s dF_1(s)$$

$$= 1+a+b - \int\limits_{1+a-b}^{1+a+b} F_1(s)ds$$

by partial integration, we can conlude

$$\gamma^*(\tilde{T}(F_\theta)) = \sup_{x\in(0,\infty)} |\theta \cdot IF(x;F_1,\tilde{T})| = \theta \cdot \frac{1+a+b-(1+a)d}{\int\limits_{1+a-b}^{1+a+b} s\, dF_1(s)}$$

$$= \theta \cdot \left| \frac{1+a+b-(1+a+b) + \int\limits_{1+a-b}^{1+a+b} F_1(s)\, ds}{\int\limits_{1+a-b}^{1+a+b} s\, dF_1(s)} \right| = \theta \cdot \frac{\int\limits_{1+a-b}^{1+a+b} F_1(s)\, ds}{\int\limits_{1+a-b}^{1+a+b} s\, dF_1(s)}.$$

In addition

$$\int\limits_{1+a-b}^{1+a+b} F_1(s)ds < \int\limits_{1+a-b}^{1+a+b} ds = 2b$$

which means $\gamma^\star(\tilde{T}(F_\theta)) < 2\gamma^\star(T(F_\theta))$.

4.3 Finite Sample Breakdown Points and an Optimal Choice for b

We consider a random sample of size n from data model (1.10). The finite sample breakdown point $\varepsilon_n^\star$ was defined in (2.13)

$$\varepsilon_n^\star(F_n, T) = \frac{1}{n} \min \left\{ m : T(\hat{F}_m) = \infty \text{ or } T(\hat{F}_m) = 0 \right\}.$$

Recall from Section 3.1, formula (3.6)

$$T_{n,i} = \frac{1}{1+a-\frac{l_1}{n}(1+a-b)-\frac{l_2}{n}(1+a+b)} \cdot \frac{1}{n} \sum_{j=l_1+1}^{n-l_2} X_{i(j)}.$$

We can assume $1 + a > 0$, because for $1 + a = 0$,

$$\sum_{j=1}^{n} \psi \left(\frac{X_{ij}}{T_{n,i}} \right) = \sum_{j=1}^{n} \min \left\{ \frac{X_{ij}}{T_{n,i}}, b \right\} > 0,$$

yielding no solution to (3.4).

Although actually $l_1 = l_1(T_{n,i})$ and $l_2 = l_2(T_{n,i})$, i.e. both indices depend on the value of $T_{n,i}$, we have

$$0 \leq l_1, l_2 \leq n,$$

which means that the denominator of $T_{n,i}$ is non-zero in any case. Now let l_u be the index such that

$$x_{i(n-l_u+1)} = x_{i(n-l_u+2)} = \ldots = x_{i(n)} = y$$

then

$$\lim_{y \to \infty} \sum_{j=1}^{n} x_{i(j)} = \lim_{y \to \infty} \left(\sum_{j=1}^{n-l_u} x_{i(j)} + l_u \cdot y \right) = \infty$$

yielding $\lim_{y \to \infty} T_{n,i}(\mathbf{x}) = \infty$ as well.

On the other hand

$$\lim_{y \to \infty} \sum_{j=1}^{n-l_u} x_{i(j)} < \infty,$$

i.e. $T_{n,i}(\mathbf{x})$ is finite. Thus a reasonable choice for l_2 would be $l_2 = l_u$. For l_ℓ being the index such that

$$x_{i(1)} = x_{i(2)} = \ldots = x_{i(l_\ell)} = z < x_{i(l_\ell+1)} \leq \ldots \leq x_{i(n)}$$

we get

$$\lim_{z \to 0} \sum_{j=1}^{n} x_{i(j)} = \lim_{z \to 0} \left(l_\ell \cdot z + \sum_{j=l_\ell+1}^{n} x_{i(j)} \right) = 0 \iff l_\ell = n.$$

According to (2.13)

$$\varepsilon_n^\star(F_n, T) = \frac{1}{n} \min \{l_2, n\} = \frac{l_2}{n}.$$

To find the finite sample breakdown point of $\tilde{T}_{n,i}$ we adapt the approach of Gather [Gather, Schultze 1999]. Because of (3.9)

$$\tilde{T}_{n,i} = \frac{1}{(1+a)d - (1+a-b)\frac{l_1}{n} - (1+a+b)\frac{l_2}{n}} \cdot \frac{1}{n} \sum_{j=l_1+1}^{n-l_2} X_{i(j)} = \sum_{j=1}^{n} w_j X_{i(j)},$$

where

$$
w_j = \begin{cases} 0 & j \in \{1, \ldots, l_1, \, n - l_2 + 1, \ldots, n\} \\ (1+a)d - (1+a-b)\frac{\tilde{l}_1}{n} - (1+a+b)\frac{\tilde{l}_2}{n} \cdot \frac{1}{n} & j = l_1 + 1, \ldots, n - l_2. \end{cases}
$$

Let $l < l_2 - 1$ and let $x_{i(n-l)} = \ldots = x_{i(n)} = y$, then

$$
\lim_{y \to \infty} T(F_n(\cdot; x_{i(1)}, \ldots, x_{i(n-l-1)}, y, \ldots, y)) = T_{n,i}(\mathbf{x}) < \infty
$$

since $w_{n-l} = \ldots = w_n = 0$.
On the other hand, if $l = l_2$ and $x_{i(n-l)} = \ldots = x_{i(n)} = y$, then

$$
\lim_{y \to \infty} T(F_n(\cdot; x_{i(1)}, \ldots, x_{i(n-l-1)}, y, \ldots, y)) = \lim_{y \to \infty} \left(\sum_{j=l_1+1}^{n-l_2-1} w_j x_{i(j)} + w_{n-l_2} y \right) = \infty.
$$

We also recognise that $\tilde{T}_{n,i}(\mathbf{x}) > 0$ as long as $x_{(n)} > 0$.
That means for the finite sample breakdown point we get

$$
\varepsilon_n^\star(F_n, \tilde{T}) = \frac{l_2}{n} \quad \text{and} \quad \lim_{n \to \infty} \varepsilon_n^\star(F_n, \tilde{T}) = \lim_{n \to \infty} \frac{l_2}{n} = 1 - F_1(1 + a + b),
$$

compare to page 50.
Refering to Section 2.3 the optimal finite sample breakdown point is $1/2$. Therefore it is reasonable to choose a, b such that

$$
1 - F_1(1 + a + b) = \frac{1}{2}.
$$

We can determine b for the asymptotic values of a from Lemma 3.8 and Lemma 3.12

$$
a_{\text{asym}} = 0 : \qquad b = \ln 2 - \alpha
$$
$$
a_{\text{asym}} = b - 1 : \qquad b = \frac{1}{2} \ln 2.
$$

Note that $b = \ln 2 - 1 < 0$ is not permitted by Lemma 2.10. Therefore, the finite sample breakdown point of $T_{n,i}$ is always less than optimal.

In the last chapter of this thesis we examine the behaviour of our estimators running some simulations. Besides we go back to the original problem of determining a credibility insurance premium.

Chapter 5

Estimation and Simulation

During the last chapters we developed two estimators $T_{n,i}$ and $\tilde{T}_{n,i}$ for risk parameters θ_i, $i = 1, \ldots, N$, that are now going to be used to solve our original problem of determining a reasonable insurance premium. Again we restrict ourselves to the Exponential distribution.

The median is known as the most robust estimator with finite sample breakdown point $\varepsilon_n^\star = \lfloor \frac{n}{2} \rfloor$, cf. [Hampel et al. 1986]. Our estimation will show that for reasonable chosen b, the asymptotic estimates $T_{n,i}(\mathbf{x}; 0, b)$ and $\tilde{T}_{n,i}(\mathbf{x}; b-1, b)$ are close the the median estimate $x_{[0.5]}$. We also discuss the different behaviour of the exact and asymptotic estimates regarding different contamination levels, samples sizes and values for b.

We recall from Section 1.4 that the insurance premium $\mu_{\text{ind},i}$ for risk class i is assumed to consist of an ordinary part $\mu_{\text{ord},i} = \mu(\theta_i)$ and an excess part μ_{xs}.

And we developed the idea of determining the next insurance period's premium through

$$\hat{\mu}_{\text{rC},i} = \hat{c}_{r,i} \cdot T_{n,i} + (1 - \hat{c}_{r,i}) \cdot \mu_{T_{n,i}}, \quad \hat{c}_{r,i} = \frac{n \cdot \widehat{\text{Var}}(E(T_{n,i})|\theta_i)}{n \cdot \widehat{\text{Var}}(E(T_{n,i})|\theta_i) + \hat{E}(\text{Var}(T_{n,i}|\theta_i))}.$$

that is an empirical robust credibility premium.

This thesis closes with a simulation of an insurance portfolio consisting of 25 risk classes. Analysing the simulation results, we explain when the actuary should employ robust credibility.

5.1 Estimating $T_{n,i}$ and $\tilde{T}_{n,i}$

It is clear from both (3.6) and (3.9) that neither $T_{n,i}(\mathbf{x})$ nor $\tilde{T}_{n,i}(\mathbf{x})$ can be calculated directly from the data, due to the dependencies of the indices l_1, l_2, $\tilde{l}_1$, $\tilde{l}_2$ from $T_{n,i}$ and θ_i, respectively.

Thus we go back to the definitions of $T_{n,i}(\mathbf{x})$ and $\tilde{T}_{n,i}(\mathbf{x})$ as solutions of

$$h(t) = \sum_{j=1}^{n} \psi\left(\frac{x_{ij}}{t}\right) = 0, \quad \psi(x) = \max\left\{-b, \min\left\{x - (1+a), b\right\}\right\} \tag{5.1}$$

$$\tilde{h}(t) = \sum_{j=1}^{n} \tilde{\psi}\left(\frac{x_{ij}}{t}\right) = 0, \quad \tilde{\psi}(x) = \frac{1}{d}\max\left\{1 + a - b, \min\left\{x, 1 + a + b\right\}\right\} - (1+a) \tag{5.2}$$

It is evident that Newton's method for seeking a zero cannot be applied because h and $\tilde{h}$ are not differentiable.

Instead, we are going to apply the bisection method, which we briefly describe below. For further information we refer to Breuer, Zwas [Breuer, Zwas 1993] for example.

Let $f : [a, b] \to \mathbb{R}$ be a continuous real-valued function with $a < b$, $[a, b] \subset \mathbb{R}$. We are looking for the solution of the problem

$$f(t) = 0, \qquad t \in [a, b]. \tag{5.3}$$

To explain the approach of the bisection method, we assume f to be monotone increasing. If $f(a) \cdot f(b) < 0$, i.e. there exists a zero in (a, b), then the bisection method starts at

$$t_1 = \frac{a + b}{2}.$$

In every iteration step, the interval $I_k = [t_k^l, t_k^u]$ containing the zero is determined as follows

$$t_k = \frac{t_{k-1}^l + t_{k-1}^u}{2}$$

$$\texttt{if} \qquad f\left(t_{k-1}^l\right) \cdot f\left(t_k^l\right) > 0$$

$$t_k^l = t_k \quad \text{and} \quad t_k^u = t_{k-1}^u$$

$$\texttt{else}$$

$$t_k^l = t_{k-1}^l \quad \text{and} \quad t_k^u = t_k$$

$$\texttt{end}$$

$$I_k = \left[t_k^l, t_k^u\right] \tag{5.4}$$

with $k = 1, 2 \ldots$ and $t_0^l = a$, $t_0^u = b$.

For $x \in \mathbb{R}$ consider the function $g : (0, \infty) \to \mathbb{R}$,

$$g(t) = \begin{cases} b, & t < \frac{x}{1+a+b} \\ \frac{x}{t} - (1+a), & \frac{x}{1+a+b} \leq t \leq \frac{x}{1+a-b} \\ -b, & t > \frac{x}{1+a-b} \end{cases}$$

with $1 + a - > 0$.

Obviously g is continuous because

$$\lim_{t \to \left(\frac{x}{1+a+b}\right)+} g(t) = \frac{x}{\frac{x}{1+a+b}} - (1+a) = b = \lim_{t \to \left(\frac{x}{1+a+b}\right)-} g(t)$$

$$\lim_{t \to \left(\frac{x}{1+a-b}\right)-} g(t) = \frac{x}{\frac{x}{1+a-b}} - (1+a) = -b = \lim_{t \to \left(\frac{x}{1+a-b}\right)+} g(t)$$

and it is monotone decreasing in t with strict monotonicity in $\left[\frac{x}{1+a+b}, \frac{x}{1+a-b}\right]$.

Thus

$$h(t) = \sum_{j=1}^{n} \psi\left(\frac{x_{ij}}{t}\right)$$

is continuous and decreasing with strict monotonicity in $\left[\frac{x_{i(1)}}{1+a+b}, \frac{x_{i(n)}}{1+a-b}\right]$, $i = 1, \ldots, N$.

Besides we can conclude that there exists a unique z_0 solving (5.1), since

$$h\left(\frac{x_{i(1)}}{1+a+b}\right) = \sum_{j=1}^{n} \max\left\{-b, \min\left\{\frac{x_{ij}}{x_{i(1)}} \cdot (1+a+b) - (1+a), b\right\}\right\} = n \cdot b > 0$$

and

$$h\left(\frac{x_{i(n)}}{1+a-b}\right) = \sum_{j=1}^{n} \max\left\{-b, \min\left\{\frac{x_{ij}}{x_{i(n)}} \cdot (1+a-b) - (1+a), b\right\}\right\} = -n \cdot b < 0.$$

The uniqueness of z_0 follows immediately from the strict monotonicity in $\left[\frac{x}{1+a+b}, \frac{x}{1+a-b}\right]$.

It is interesting to note that in the case $1 + a - b < 0$

$$h(t) = \sum_{j=1}^{n} \max\left\{-b, \min\left\{\frac{x_{ij}}{t} - (1+a), b\right\}\right\}$$

$$= \sum_{j=1}^{n} \max\left\{1 + a - b, \min\left\{\frac{x_{ij}}{t}, 1 + a + b\right\}\right\} - n(1+a)$$

$$= \sum_{j=1}^{n} \min\left\{\frac{x_{ij}}{t}, 1 + a + b\right\} - n(1+a),$$

because $x_{ij} > 0$ for all $j = 1, \ldots, n$, $i = 1, \ldots, N$. In that case

$$h\left(\frac{x_{i(n)}}{1+a}\right) = \sum_{j=1}^{n} \min\left\{\frac{x_{ij}}{x_{i(n)}} \cdot (1+a), 1 + a + b\right\} - n(1+a)$$

$$< n \cdot \min\{1 + a, 1 + a + b\} - n(1+a) = 0$$

yielding that the zero z_0 lies in $\left[\frac{x_{i(1)}}{1+a+b}, \frac{x_{i(n)}}{1+a}\right]$.

In the same manner, considering

$$\tilde{g}(t) = \frac{1}{d}\begin{cases} 1+a+b-(1+a)d, & t < \frac{x}{1+a+b} \\ \frac{x}{t}-(1+a)d, & \frac{x}{1+a+b} \leq t \leq \frac{x}{1+a-b} \\ 1+a-b-(1+a)d, & t > \frac{x}{1+a-b}, \end{cases}$$

we get the continuity and strict monotonicity of $\tilde{h}$ in (5.2). Because

$$\tilde{h}\left(\frac{x_{i(1)}}{1+a+b}\right)$$

$$= \frac{1}{d}\sum_{j=1}^{n}\max\left\{1+a-b, \min\left\{\frac{x_{ij}}{x_{i(1)}}\cdot(1+a+b), 1+a+b\right\}\right\} - n\cdot(1+a)$$

$$= \frac{n}{d}\cdot(1+a+b-(1+a)d) = \frac{n}{d}\cdot\int_{1+a-b}^{1+a+b}F_1(s)\,ds > 0$$

due to the definition of $(\alpha+a)d = (1+a)d$ in equation (3.10) and

$$\tilde{h}\left(\frac{x_{i(n)}}{1+a-b}\right)$$

$$= \frac{1}{d}\sum_{j=1}^{n}\max\left\{1+a-b, \min\left\{\frac{x_{ij}}{x_{i(n)}}\cdot(1+a-b), 1+a+b\right\}\right\} - n\cdot(1+a)$$

$$= \frac{n}{d}\cdot(1+a-b-(1+a)d)$$

$$= \frac{n}{d}\cdot\left(-2b + \int_{1+a-b}^{1+a+b}F_1(s)\,ds\right) < 0,$$

due to $F_1(s) < 1$, we conclude that the zero $\tilde{z}_0$ of (5.2) lies in $\left[\frac{x_{i(1)}}{1+a+b}, \frac{x_{i(n)}}{1+a-b}\right]$.

Here, the crucial case is $1+a-b=0$, i.e. $a = a_{\mathrm{asym}} = b-1$. In all other cases $a > b-1$, cf. proof of Theorem 3.11. If $1+a-b=0$, then $z_0 \in \left[\frac{x_{i(1)}}{1+a+b}, \infty\right)$.

Knowing all this, we start the computation of $T_{n,i}(\mathbf{x}; a, b)$ and $\tilde{T}_{n,i}(\mathbf{x}; a, b)$ evaluating $h, \tilde{h}$ at

$$t_0 = \frac{x_{i(1)}}{1+a+b}$$

$$t_1 = x_{i(1)}$$

$$\cdots$$

$$t_l = x_{i(l)}$$

$$t_{l+1} = \frac{x_{i(n)}}{1+a-b}$$

with index l chosen such that

$$x_{(l)} \leq \frac{x_{i(n)}}{1 + a - b} < x_{(l+1)}.$$

Because of the above considerations, there will be indices k, $\tilde{k}$ with

$$h(t_k) > 0, \qquad h(t_{k+1}) < 0$$
$$\tilde{h}\left(t_{\tilde{k}}\right) > 0, \qquad \tilde{h}\left(t_{\tilde{k}+1}\right) < 0.$$

So, the intervals to begin our bisection routines are $[t_k, t_{k+1}]$ and $[\tilde{t}_{\tilde{k}}, \tilde{t}_{\tilde{k}+1}]$.

But instead of starting the bisection calculation routine directly, we install one more step beforehand. From the definitions of h and $\tilde{h}$, we realise, that both functionals are either linear or at least stepwise linear in the intervals $[t_k, t_{k+1}]$, $k = 1, \ldots, l$.
In case the function is linear, we can give the zeros $z_0 \in (t_k, t_{k+1})$ and $\tilde{z}_0 \in (t_{\tilde{k}}, t_{\tilde{k}+1})$ directly

$$z_0 = t_k - \frac{h(t_k)}{h(t_{k+1}) - h(t_k)} \cdot (t_{k+1} - t_k)$$
$$\tilde{z}_0 = t_k - \frac{h(t_{\tilde{k}})}{h(t_{\tilde{k}+1}) - h(t_{\tilde{k}})} \cdot \left(t_{\tilde{k}+1} - t_{\tilde{k}}\right).$$

Thus, the calculation routine is

$$I_0 = [t_k, t_{k+1}], \qquad \texttt{where} \quad h(t_k) \cdot h(t_{k+1}) < 0$$
$$z_0 = t_k - \frac{h(t_k)}{h(t_{k+1}) - h(t_k)} \cdot (t_{k+1} - t_k)$$

if $h(z_0) < \varepsilon$

    STOP

else

    if $h(t_k) \cdot h(z_0) < 0$

        $I_k = [t_k, z_0]$

    else

        $I_k = [z_0, t_{k+1}]$

    end

    bisection (5.4)

end

with error bound ε. As stop criterion in the bisection routine we use $h(t_k) < \varepsilon$ as well. The same algorithm is applied for $\tilde{h}$ and $\tilde{k}$ instead of h and k.

We explained in Section 1.4 that the Gamma distribution $\Gamma(\alpha, \theta)$ is commonly used in insurance mathematics, because it naturally arises in the theory of homogeneous Poisson processes, as for example explained by Mikosch [Mikosch 2004]. Besides the Gamma distribution is closed under summation, cf. Johnson et al. [Johnson et al. 1994], which means that the sum of Gamma distributed claim amounts is again Gamma distributed, under minor restrictions regarding the parameters.

Likewise we noted in Section 1.4 that the Pareto distribution is often used when the actuary has to deal with large claims. This is discussed by Rolski et al. [Rolski et al. 1999].

Therefore we chosed our data model (1.10) in Chapter 1

$$\mathcal{F}_\varepsilon(\Gamma(\alpha, \theta)) = \{F : F = (1 - \varepsilon) \cdot \Gamma(\alpha, \theta) + \varepsilon \cdot Par(\lambda, x_0)\}.$$

The parametrisation of the Gamma distribution is chosen such that

$$f_\Gamma(x) = \frac{1}{\Gamma(\alpha) \cdot \theta^\alpha} \cdot x^{\alpha-1} \cdot e^{-x/\theta}, \qquad x > 0$$

is the density. Then $E_\Gamma(X) = \alpha \cdot \theta$ for $X \sim \Gamma(\alpha, \theta)$. The density of the Pareto distribution $Par(\lambda, x_0)$ is given by

$$f_{\text{Par}}(x) = \frac{\lambda \cdot x_0^\lambda}{x^{\lambda+1}}, \qquad x > x_0, \, x_0, \lambda > 0$$

with expected value $E_{\text{Par}}(X) = (\lambda \cdot x_0)/(\lambda - 1)$.

We choose $\alpha = 1$, $\theta = 3$ and $\lambda = 2$ for the calculations below. Note that the Pareto distribution function has support (x_0, ∞).

To model the large claims, i.e. the contamination of $\Gamma(1, 3) = Exp(3)$, we are going to consider two different Pareto distributions. In the first case, we assume the minimum excess-claim x_0 of the Pareto distribution to be 5 times as much as the expected claim amount of the $\Gamma(1, 3)$-distribution; meaning we have $x_0 = 3 \cdot 5 = 15$ and the mean excess-claim amount of the Pareto-distribution is $E_{\text{Par}}(X) = 30$.

The second case considers a minimum excess-claim amount x_0 of as much as 50 times the expected ordinary claim amount, that is $x_0 = 3 \cdot 50 = 150$ and the expected value of the excess claim is $E_{\text{Par}}(X) = 300$.

The random numbers used for all calculations to follow were generated in R applying the functions rgamma() to gain random deviates from the Gamma distribution. To get Pareto random numbers, the function urpareto() of the Runuran package by Leynold and Hörmann was applied.

Both functions use Uniform-$[0, 1]$ random numbers generated from the Mersenne Twister TM 19937. The transformation in Γ-random numbers is done according to a rejection technique described by Ahrens, Dieter [Ahrens, Dieter 1982]. The generation algorithm in urpareto() uses fast numerical inversion.

To get a picture of the estimators' behaviour we draw sample sizes of different lengths. Besides we consider several contamination degrees ε. Additionally we examine the behaviour of $T_{n,i}(a, b)$ and $\tilde{T}_{n,i}(a, b)$ with respect to different values of b.

The values for a are chosen according to

- the Fisher approach, cf. formula (3.16); then $a = a_{\text{Fish}}$

- Lemma 3.8, the asymptotic value of the Fisher approach; then $a = 0$

- the Minimum Variance approach, cf. formula (3.26), that is

$$\left(e^{2b} - 1\right)^3 + e^{1+a+b}\left(e^{2b} - (1+2b)\right)\left(e^{2b}(2+a-b) - (2+a+b) - 2(e^{2b}-1)\right) = 0;$$

then $a = a_{\text{MV}}$

- Lemma 3.12, the asymptotic value of the Minimum Variance approach; then $a = b - 1$.

Below, the graphs show different estimates for sample sizes $n \in \{5, 10, 30, 100, 500, 100\}$ and contamination degress $\varepsilon \in \{0.001, 0.1, 0.5\}$. The parameter $k \in \{5, 50\}$ denotes the factor that is used to determine x_0 in $Par(\lambda, x_0)$. For b we choose a small value $b = 0.01$, the optimal value $b = 1/2 \cdot \ln 2$ to get the best possible finite sample breakdown point, cf. Section 4.3, and a comparatively bigger value $b = 4$ that assures the values a_{Fish} and a_{MV} to be close to their asymptotic values 0 and $b - 1$.

For reasons of better reading, we go back to the notations of Sections 3.3.1 and 3.3.2, that is $T_n(\mathbf{x}; a, b)$ and $\tilde{T}_n(\mathbf{x}; a, b)$. Labelling the graphs, we omit the tilde sign since it is clear that a_{MV} and $a = b - 1$ belong to $\tilde{T}_n$.
For comparison we give both the mean value and the median for each sample as well.

Recall again that we only considered one random sample in each case. But for addressing the analysis of the four estimators' behaviour in different situations, this is just what we need.
We see immediately that for $b = 4$, the estimates are close together, often even close to the mean $\bar{X}$, cf. Figures 5.3, 5.6, 5.9, 5.12, 5.15, 5.18.
To explain this behaviour, we have a look at formulae (3.4) and (3.11) again

$$[(3.4)]: \quad \sum_{j=1}^{n} \psi\left(\frac{x_{ij}}{T_{n,i}(\mathbf{x}; 0, b)}\right) = 0, \quad \psi(x) = \max\left\{-b, \min\left\{x - (1+a), b\right\}\right\}$$

$$[(3.11)]: \quad \sum_{j-1}^{n} \tilde{\psi}\left(\frac{x_{ij}}{\tilde{T}_{n,i}(\mathbf{x}; b-1, b)}\right) = 0, \quad \tilde{\psi}(x) = \tfrac{1}{d}\left(\psi(x) + (1+a)\right) - (1+a).$$

Note that for bigger b (here $b = 4$), both a_{Fish} and a_{MV} are close to their asymptotic values $a_{\text{Fish,asym}} = 0$ and $a_{\text{MV,asym}} = b - 1$, respectively. And for the asymptotic values a_{Fish} and a_{MV} we get

$$1 + a_{\text{Fish,asym}} - b = 1 - b < 0, \quad 1 + a_{\text{MV,asym}} - b = 0.$$

Estimates for $\varepsilon = 0.001$, $k = 5$ and different values for b

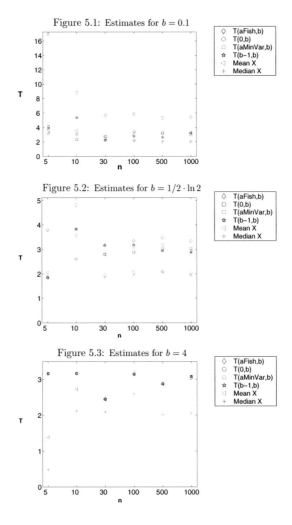

Figure 5.1: Estimates for $b = 0.1$

Figure 5.2: Estimates for $b = 1/2 \cdot \ln 2$

Figure 5.3: Estimates for $b = 4$

Estimates for scale parameter θ in ε-contamination model
$0.999 \cdot \Gamma(1,3) + 0.001 \cdot Par(2,15)$.

Estimates for $\varepsilon = 0.001$, $k = 50$ and different values for b

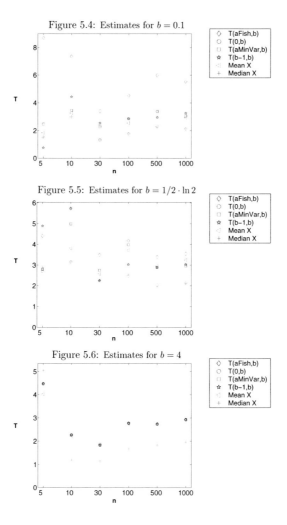

Estimates for scale parameter θ in ε-contamination model
$0.999 \cdot \Gamma(1,3) + 0.001 \cdot Par(2,150)$.

Estimates for $\varepsilon = 0.1$, $k = 5$ and different values for b

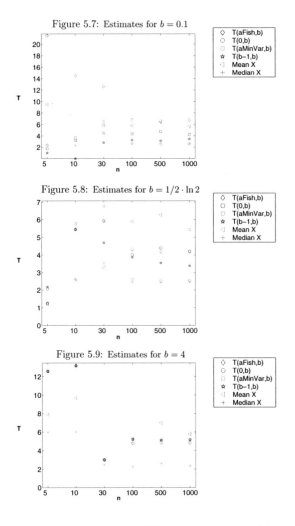

Estimates for scale parameter θ in ε-contamination model
$0.9 \cdot \Gamma(1,3) + 0.1 \cdot Par(2,15)$.

Estimates for $\varepsilon = 0.1$, $k = 50$ and different values for b

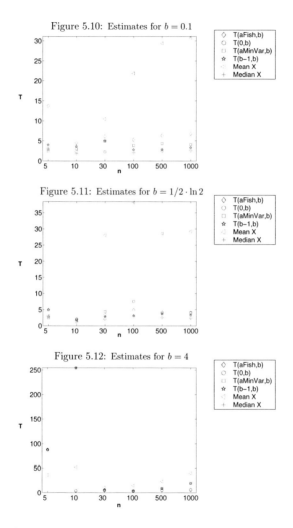

Estimates for scale parameter θ in ε-contamination model
$0.9 \cdot \Gamma(1,3) + 0.1 \cdot Par(2,150)$.

Estimates for $\varepsilon = 0.5$, $k = 5$ and different values for b

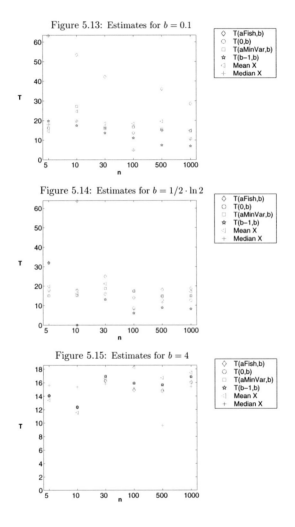

Estimates for scale parameter θ in ε-contamination model
$0.5 \cdot \Gamma(1,3) + 0.5 \cdot Par(2,15)$.

Estimates for $\varepsilon = 0.5$, $k = 50$ and different values for b

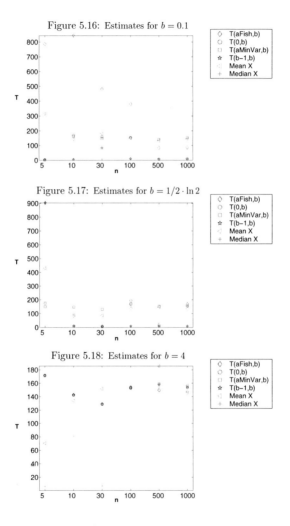

Figure 5.16: Estimates for $b = 0.1$

Figure 5.17: Estimates for $b = 1/2 \cdot \ln 2$

Figure 5.18: Estimates for $b = 4$

Estimates for scale parameter θ in ε-contamination model
$0.5 \cdot \Gamma(1,3) + 0.5 \cdot Par(2, 150)$.

Thus for $b > 1$

$$\sum_{j=1}^{n} \max \left\{ -b, \min \left\{ \frac{x_{ij}}{T_{n,i}(\mathbf{x}; a_{\text{Fish}}, b)} - (1 + a_{\text{Fish}}), b \right\} \right\}$$

$$= \sum_{j=1}^{n} \min \left\{ \frac{x_{ij}}{T_{n,i}(\mathbf{x}; a_{\text{Fish}}, b)} - (1 + a_{\text{Fish}}), b \right\}$$

because $x_{ij}/T_{n,i}(\mathbf{x}; a_{\text{Fish}}, b) > 0$ for all $j = 1, \ldots, n$, $i = 1, \ldots, N$.
Furthermore, we point out that the arithmetic mean $\bar{X}_i$ is the solution to

$$\sum_{j=1}^{n} \left(\frac{X_{ij}}{\theta_i} - \alpha \right) = \sum_{j=1}^{n} \left(\frac{X_{ij}}{\theta_i} - 1 \right) = 0.$$

For a random variable X with distribution $\Gamma(1,3)$, hence $E(X) = 3$ and $median(X) = (\ln 2)/3$, it is less likely to take values much bigger than 4. Indeed, the probablity is $1 - F_3(4) = e^{-4}$.
Therefore, if the contamination level ε of the random sample $X_{i1}, \ldots, X_{in}$ is small, the estimates $T_{n,i}(\mathbf{x}; 0, b)$ will be close to $\bar{x}$.

The same kind of performance can be observed for $\tilde{T}_{n,i}$. Especially, the estimate $\tilde{T}_{n,i}(x; b-1, b)$ with asymptotic $a_{\text{MV,asym}} = b - 1$ lies close to $\bar{x}$. We can explain this noting that for $a_{\text{MV,asym}}$ we get

$$\sum_{j=1}^{n} \tilde{\psi} \left(\frac{x_{ij}}{\tilde{T}_{n,i}(\mathbf{x}; b-1, b)} \right)$$

$$= \sum_{j=1}^{n} \left[\frac{1}{d} \left(\max \left\{ -b, \min \left\{ \frac{x_{ij}}{\tilde{T}_{n,i}(\mathbf{x}; b-1, b)} - (1 + b - 1), \right\} \right\} + (1 + b - 1) \right) \right.$$

$$\left. - (1 + b - 1) \right]$$

$$= \frac{1}{d} \sum_{j=1}^{n} \left(\min \left\{ \frac{x_{ij}}{\tilde{T}_{n,i}(\mathbf{x}; b-1, b)}, 2b \right\} - (1 + b - 1)d \right)$$

because $x_{ij}/\tilde{T}_{n,i}(\mathbf{x}; b-1, b) > 0$ for all $j = 1, \ldots, n$, $i = 1, \ldots, N$. Since $F_3 = \Gamma(1,3)$ is light-tailed, we get for $b = 4$

$$(1 + a)d = (1 + b - 1)d$$

$$= 2b \cdot (1 - F_3(2b)) - \int_0^{2b} s dF_3(s)$$

$$= 8e^{-8} - (1 - 8e^{-8}) = -1 + 16e^{-8}.$$

Again $\bar{x}$ solves

$$\sum_{j=1}^{n} \left(\frac{x_{ij}}{\theta_i} - 1 \right) = 0.$$

Hence also for less contaminated samples, the estimate $\tilde{T}_n(\mathbf{x}; b-1, b)$ will be close to $\bar{x}$.

For $b = 0.1$ we notice something else, cf. Figures 5.1, 5.4, 5.7, 5.10, 5.13, 5.16. In some cases the robust estimates lie above the arithmetic mean. Again, this especially occurs if the contamination – either in terms of ε or k – is comparatively small. To better understand this behaviour, we rewrite $T_{n,i}(\mathbf{x}; a_{\mathrm{Fish}}, b)$

$$T_{n,i}(\mathbf{x}; a_{\mathrm{Fish}}, b) = \frac{\frac{1}{n}\sum\limits_{j=l_1}^{n-l_2} x_{i(j)}}{(1 + a_{\mathrm{Fish}}) - \frac{l_1}{n}(1 + a_{\mathrm{Fish}} - b) - \frac{l_2}{n}(1 + a_{\mathrm{Fish}} + b)}$$

$$= \frac{\bar{x}_i - \frac{1}{n}\sum\limits_{j=1}^{l_1-1} x_{i(j)} - \frac{1}{n}\sum\limits_{j=n-l_2+1}^{n} x_{i(j)}}{(1 + a_{\mathrm{Fish}}) - \frac{l_1}{n}(1 + a_{\mathrm{Fish}} - b) - \frac{l_2}{n}(1 + a_{\mathrm{Fish}} + b)}.$$

In more homogeneous samples, i.e. samples with no or only small contamination, the indices l_1 and l_2 will be small as well as the second and third summand of the numerator. Since $0 < 1 + a_{\mathrm{Fish}} < 1$, this may lead to $T_{n,i}(\mathbf{x}; a_{\mathrm{Fish}}, b) > \bar{x}_i$.

Rewriting $\tilde{T}_{n,i}(\mathbf{x}; a_{\mathrm{MV}}, b)$ in the same manner

$$\tilde{T}_{n,i}(\mathbf{x}; a_{\mathrm{MV}}, b) = \frac{\frac{1}{n}\sum\limits_{j=\tilde{l}_1}^{n-\tilde{l}_2} x_{i(j)}}{(1 + a_{\mathrm{MV}})d - \frac{\tilde{l}_1}{n}(1 + a_{\mathrm{MV}} - b) - \frac{\tilde{l}_2}{n}(1 + a_{\mathrm{MV}} + b)}$$

$$= \frac{\bar{x}_i - \frac{1}{n}\sum\limits_{j=1}^{\tilde{l}_1-1} x_{i(j)} - \frac{1}{n}\sum\limits_{j=n-\tilde{l}_2+1}^{n} x_{i(j)}}{(1 + a_{\mathrm{MV}})d - \frac{\tilde{l}_1}{n}(1 + a_{\mathrm{MV}} - b) - \frac{\tilde{l}_2}{n}(1 + a_{\mathrm{MV}} + b)},$$

we realise that $\tilde{T}_{n,i}(\mathbf{x}; a_{\mathrm{MV}}, b)$ might be bigger than $\bar{x}_i$ as well for the same reasons as $T_{n,i}(\mathbf{x}; a_{\mathrm{Fish}}, b)$.

5.2   Simulation Study

Now that we have seen, how the estimators behave in certain situations, we want to judge the quality of their estimation. In this section we will omit the index i of the risk class. Even though the following considerations only deal with T_n, they also hold for $\tilde{T}_{n,\cdot}$.

A possibility to check the estimators' quality is the mean square deviation $E(||T_n - \theta||^2)$

(cf. Lehn, Wegmann [Lehn, Wegmann 2004])

$$E\left(||T_n - \theta||^2\right)$$
$$= E\big[T_n - E\left(T_n\right) + E\left(T_n\right) - \theta\big]^2$$
$$= E\big[T_n - E\left(T_n\right)\big]^2 + 2\big[E\left(T_n\right) - \theta\big] \cdot \underbrace{E\big[T_n - E\left(T_n\right)\big]}_{=0} + E\big[E\left(T_n\right) - \theta\big]^2$$
$$= \mathrm{Var}\left(T_n\right) + E\big[E\left(T_n\right) - \theta\big]^2. \tag{5.5}$$

Thus computing $\mathrm{Var}(T_n)$ in addition to $E\left(||T_n - \theta||^2\right)$ we also get information on the bias $E(T_n - \theta)$ of the estimator.

The Monte Carlo study is carried out as follows. Estimating $E\left(||T_n - \theta||^2\right)$ we use

$$\hat{d}_n := \hat{E}\left(||T_n - \theta||^2\right) = \frac{1}{K}\sum_{k=1}^{K}\left(T_n^k - \theta\right)^2,$$

where K is the number of Monte Carlo steps and T_n^k is the estimator for the k-th random sample. The unbiased variance estimator

$$\hat{v}_n := \widehat{\mathrm{Var}}(T_n) = \frac{1}{K-1}\sum_{k=1}^{K}\left(T_n^k - \bar{T}\right)^2, \qquad \bar{T} = \frac{1}{K}\sum_{k=1}^{K}T_n^k$$

is used to estimate $\mathrm{Var}(T_n)$.

Below we show in Tables 5.1 and 5.2 the estimates for the mean square deviations and the variances of both T_n and $\tilde{T}_n$ for different sample sizes n, contamination degrees ε and different values for θ. We also give the estimates of the mean $\bar{X}$ and the median $X_{[0.5]}$.

The first thing we note is that the estimate $T_n(\mathbf{x}; a, b)$ is always bigger than its asymptotic counterpart $T_n(\mathbf{x}; 0, b)$. Even though this seems to be a surprising result, there is an analytical explanation. Recall definition (3.4) of estimator T_n. From Section 3.3.1 we know that $a_{\mathrm{Fish}} < 0$. Thus

$$\frac{x_j}{t} - (1 + a_{\mathrm{Fish}}) > \frac{x_j}{t} - 1, \qquad j = 1, \ldots, n,$$

which yields

$$\sum_{j=1}^{n}\max\left\{-b, \min\left\{\frac{x_j}{t} - (1 + a_{\mathrm{Fish}}), b\right\}\right\} \geq \sum_{j=1}^{n}\max\left\{-b, \min\left\{\frac{x_j}{t} - 1, b\right\}\right\}.$$

Thus for $T_n(\mathbf{x}; a_{\mathrm{Fish}}, b)$ and $T_n(\mathbf{x}; 0, b)$ solving

$$\sum_{j=1}^{n}\max\left\{-b, \min\left\{\frac{x_j}{T_n(\mathbf{x}; a_{\mathrm{Fish}}, b)} - (1 + a_{\mathrm{Fish}}), b\right\}\right\} = 0$$

and

$$\sum_{j=1}^{n} \max\left\{ -b, \min\left\{ \frac{x_j}{T_n(\mathbf{x}; 0, b)} - 1, b \right\} \right\} = 0$$

we get $T_n(\mathbf{x}; a_{\text{Fish}}, b) > T_n(\mathbf{x}; 0, b)$ since h in (5.1) is decreasing in t.

It is further remarkable that for $b = 4$, the estimates $\tilde{T}_n(\mathbf{x}; a_{\text{MV}}, b)$ and $\tilde{T}_n(\mathbf{x}; b - 1, b)$ are much bigger than both the mean value $\bar{x}$ and the robust estimates $T_n(\mathbf{x}; a_{\text{Fish}}, b)$, $T_n(\mathbf{x}; 0, b)$.

Looking for an explanation, we note that for $b = 4$ the exact value a_{MV} is close to the asymptotic value $b - 1$, i.e. $1 + a_{\text{MV}} - b$ is close to 0, whereas $1 + (b - 1) - b = 0$. On the other hand $1 + a_{\text{MV}} + b$ is comparatively big yielding

$$\min\left\{ \frac{x_j}{\tilde{T}_n(\mathbf{x}; a_{\text{MV}})}, 1 + a + b \right\} = \frac{x_j}{\tilde{T}_n(\mathbf{x}; a_{\text{MV}}, b)}$$

for most of the indices $j = 1, \ldots, n$.
Now, $\tilde{T}_n(\mathbf{x}; a_{\text{MV}}, b)$ solves

$$\sum_{j=1}^{n} \max\left\{ 1 + a - b, \min\left\{ \frac{x_j}{\tilde{T}_n(\mathbf{x}; a_{\text{MV}}, b)}, 1 + a_{\text{MV}} + b \right\} \right\} - n(1 + a)d = 0$$

which is close to the solution of

$$\sum_{j=1}^{n} \frac{x_j}{T_n(\mathbf{x}; a_{\text{MV}}, b)} - n(1 + a_{\text{MV}})d = 0$$

due to the above considerations. Obviously, the solution to the latter is

$$T_n(\mathbf{x}; a_{\text{MV}}) = \frac{1}{(1 + a_{\text{MV}})d} \cdot \bar{x}.$$

Because

$$(1 + a_{\text{MV}})d = (1 + a_{\text{MV}} - b) \cdot F_1(1 + a_{\text{MV}} - b) + (1 + a_{\text{MV}} + b) \cdot (1 - F_1(1 + a_{\text{MV}} + b))$$

$$+ \int_{1 \, | \, a_{\text{MV}} \, b}^{1 \, | \, a_{\text{MV}} \, | \, b} s \, dF_1(s)$$

$$= 1 + a_{\text{MV}} - b + e^{-(1 + a_{\text{MV}} - b)} - e^{-(1 + a_{\text{MV}} + b)}$$

which is certainly positive since $a_{\text{MV}} > b - 1$ and $1 + a_{\text{MV}} - b < 1 + a + b$. For $a_{\text{MV,asym}} = b - 1$ we have $(1 + a_{\text{asym}})d = 1 - e^{-2b} < 1$, which means

$$\frac{1}{(1 + a_{\text{MV,asym}})d} \cdot \bar{x} > \bar{x}.$$

		$\varepsilon = 0.001$								
		$\theta = 1$			$\theta = 3$			$\theta = 6$		
		$\bar{T}_n$	$\hat{E}(\lVert T_n - \theta \rVert^2)$	$\widehat{\mathrm{Var}}(T_n)$	$\bar{T}_n$	$\hat{E}(\lVert T_n - \theta \rVert^2)$	$\widehat{\mathrm{Var}}(T_n)$	$\bar{T}_n$	$\hat{E}(\lVert T_n - \theta \rVert^2)$	$\widehat{\mathrm{Var}}(T_n)$
$b = 0.1$	$T_n(a,b)$	1.86	0.80	0.07	5.58	7.24	0.59	11.16	28.96	2.35
	$T_n(0,b)$	0.71	0.10	0.01	2.12	0.87	0.09	4.23	3.48	0.35
	$\tilde{T}_n(a,b)$	1.01	0.02	0.02	3.04	0.14	0.14	6.09	0.57	0.56
	$\tilde{T}_n(b-1,\tilde{b})$	1.04	0.08	0.08	3.13	0.69	0.68	6.26	2.78	2.71
	$\bar{X}$	1.10	0.38	0.37	3.30	3.43	3.34	6.60	13.73	13.38
	$X_{[0.5]}$	0.70	0.10	0.01	2.10	0.90	0.09	4.19	3.62	0.36
$b = \frac{1}{2} \cdot \ln 2$	$T_n(a,b)$	1.18	0.06	0.02	3.54	0.50	0.21	7.09	2.01	0.82
	$T_n(0,b)$	0.71	0.09	0.01	2.14	0.81	0.08	4.29	3.26	0.33
	$\tilde{T}_n(a,b)$	1.01	0.01	0.01	3.04	0.13	0.13	6.08	0.52	0.52
	$\tilde{T}_n(b-1,b)$	1.02	0.03	0.02	3.06	0.23	0.22	6.12	0.91	0.89
	$\bar{X}$	1.10	0.38	0.37	3.30	3.43	3.34	6.60	13.73	13.38
	$X_{[0.5]}$	0.70	0.10	0.10	2.10	0.90	0.09	4.19	3.62	0.36
$b = 4$	$T_n(a,b)$	1.01	0.01	0.01	3.04	0.10	0.10	6.09	0.39	0.39
	$T_n(0,b)$	1.01	0.01	0.01	3.02	0.10	0.10	6.04	0.38	0.38
	$\tilde{T}_n(a,b)$	1.01	0.01	0.01	3.05	0.10	0.10	6.10	0.41	0.40
	$\tilde{T}_n(b-1,b)$	1.02	0.01	0.01	3.05	0.10	0.10	6.10	0.41	0.40
	$\bar{X}$	1.10	0.42	0.41	3.30	3.43	3.34	6.60	18.58	18.22
	$X_{[0.5]}$	0.70	0.10	0.01	2.10	0.90	0.09	4.20	3.61	0.36

Table 5.1: Results Monte-Carlo-Simulation, $\varepsilon = 0.001$, $n = 100$, $K = 600000$, $k = 50$

		$\theta = 1$			$\theta = 3$			$\theta = 6$		
		$\bar{T}_n$	$\hat{E}(\|T_n - \theta\|^2)$	$\widehat{\mathrm{Var}}(T_n)$	$\bar{T}_n$	$\hat{E}(\|T_n - \theta\|^2)$	$\widehat{\mathrm{Var}}(T_n)$	$\bar{T}_n$	$\hat{E}(\|T_n - \theta\|^2)$	$\widehat{\mathrm{Var}}(T_n)$
$b = 0.1$	$T_n(a,b)$	2.19	1.60	0.17	6.58	14.27	1.45	13.17	57.85	6.45
	$T_n(0,b)$	0.83	0.05	0.02	2.48	0.41	0.14	4.96	1.64	0.56
	$\tilde{T}_n(a,b)$	1.43	1.48	1.29	4.29	11.65	9.99	8.61	53.10	46.31
	$\tilde{T}_n(b-1,b)$	1.17	0.13	0.10	3.51	1.13	0.87	7.03	4.55	3.49
	$\bar{X}$	10.91	146.86	48.61	32.71	1368.61	485.97	65.48	5436.99	1898.65
	$X_{[0.5]}$	0.82	0.05	0.02	2.45	0.44	0.14	4.91	1.75	0.57
$b = \frac{1}{2} \cdot \ln 2$	$T_n(a,b)$	1.40	0.19	0.04	4.19	1.75	0.34	8.37	7.02	1.39
	$T_n(0,b)$	0.84	0.04	0.01	2.52	0.36	0.13	5.04	1.46	0.53
	$\tilde{T}_n(a,b)$	1.40	0.72	0.56	4.19	5.60	4.17	8.38	21.29	15.60
	$\tilde{T}_n(b-1,b)$	1.17	0.06	0.04	3.51	0.58	0.32	7.01	2.31	1.28
	$\bar{X}$	10.90	144.32	46.29	32.75	1405.34	520.01	65.38	4805.18	1279.07
	$X_{[0.5]}$	0.82	0.05	0.02	2.46	0.44	0.14	4.91	1.75	0.57
$b = 4$	$T_n(a,b)$	2.65	20.08	17.36	7.95	180.84	156.34	15.81	708.98	612.70
	$T_n(0,b)$	2.58	18.61	16.13	7.73	167.59	145.23	15.37	655.79	568.00
	$\tilde{T}_n(a,b)$	14.70	351.18	163.48	44.10	3160.91	1471.47	88.18	12643.11	5888.77
	$\tilde{T}_n(b-1,b)$	14.53	346.25	163.15	43.59	3116.26	1468.42	87.17	12466.41	5877.06
	$\bar{X}$	10.91	150.29	52.05	32.75	1405.34	520.01	65.1	5171.90	1642.94
	$X_{[0.5]}$	0.82	0.05	0.016	2.46	0.44	0.14	4.91	1.75	0.57

$\varepsilon = 0.1$

Table 5.2: Results Monte-Carlo-Simulation, $\varepsilon = 0.1$, $n = 100$, $K = 600000$, $k = 50$

Remark 5.1
We certainly have $1 + a_{Fish} + b < 1 + b < 2b < 1 + a_{MV} + b$ *if* $b > 1$. *For the same* b,
$1 + a_{Fish} - b < 0$ *and* $1 + a_{MV} - b$ *close to* 0. *That means, the random samples are not
cutted from below, but there are remarkable differences in the upper thresholds. This does
explain why for* $b = 4$, *the estimates* $\tilde{T}_n(\mathbf{x}; a_{MV}, b)$ *and* $\tilde{T}_n(\mathbf{x}; b-1, b)$ *are much bigger than*
$T_n(\mathbf{x}; a_{Fish}, b)$ *and* $T_n(\mathbf{x}; 0, b)$.

Surprisingly, we do not always get $\widehat{\mathrm{Var}}(T_n(a_{\mathrm{Fish}}, b)) < \widehat{\mathrm{Var}}(\tilde{T}_n(a_{\mathrm{MV}}, b))$, even though T_n is
chosen according to Lemma 2.10. But if we recall that Lemma 2.10 only considers the
minimal asymptotic variance, our simulation results are still consistent with the theory.

Thall [Thall 1979] mentioned, that the estimators defined by (3.4) and (3.11) are biased
because Fisher-consistency only holds at the true distribution. Our results confirm this by
the fact that both $T_n(\mathbf{x}; a_{\mathrm{Fish}}, b)$ and $\tilde{T}_n(\mathbf{x}; a_{\mathrm{MV}}, b)$ overestimate θ. It can also be detected
by considering $\hat{d}_n - \hat{v}_n$ according to (5.5).
But it seems, that choosing the asymptotic values compensates for most of the bias.

As for comparing our robust estimators with the median $X_{[0.5]}$, we see that in most cases,
the asymptotic estimates $T_n(\mathbf{x}; 0, b)$ and $\tilde{T}_n(\mathbf{x}; b-1, b)$ are close to the median value $x_{[0.5]}$.
It is especially noticable that the empirical variances of T_n and $\tilde{T}_n$ are as small as the
empirical variance of the median. And even for $\theta = 6$, i.e. a claim amount distribution
with a heavier tail, the estimates $T_n(\mathbf{x}; 0, b)$ and $\tilde{T}_n(\mathbf{x}; b-1, b)$ are quite close to $x_{[0.5]}$.
Only for the contamination level $\varepsilon = 0.1$ and $b = 4$, the median estimate outshines all
robust estimates.

5.3   Credibility Premium

We now go back to our original problem of estimating an insurance premium. Recall from
Section 1.4, that the insurance portfolio consists of N risks, each associated with a risk
parameter ϑ_i, $i = 1, \ldots, N$. The risk parameters are i.i.d. random variables with distribution (structure) function U.

For each of the N risk classes, the individual insurance premium $\mu_{\mathrm{ind},i}$ is the sum of an
ordinary part $\mu_{\mathrm{ord},i}$ and an excess part μ_{xs}, cf. (1.7). The ordinary part is assumed to be
explained by the risk parameter ϑ_i. In our model $\vartheta_i = \theta_i$.

According to (1.8), the credibility premium $\hat{\mu}_{\mathrm{C},i}$ is determined by

$$\hat{\mu}_{\mathrm{rC},i}(\theta_i) = c_{r,i} \cdot T_{n,i} + (1 - c_{r,i}) \cdot \mu_{T_{n,i}}, \qquad i = 1, \ldots, N$$

with $\mu_{T_{n,i}} = E(T_{n,i})$ and

$$c_{r,i} = \frac{n \cdot \mathrm{Var}(E(T_{n,i})|\theta_i)}{n \cdot \mathrm{Var}(E(T_{n,i})|\theta_i) + E(\mathrm{Var}(T_{n,i}|\theta_i))}, \, i = 1, \ldots, N.$$

Unfortunately – as we pointed out in Section 1.4 - we neither do know $E(\text{Var}(T_{n,i}|\theta_i))$ nor $\text{Var}(E(T_{n,i}|\theta_i))$.

That means we have to estimate them as well, leading to the *empirical robust credibility estimator*

$$\hat{\mu}_{\text{rC},i;\,n} = \hat{c}_{r,i} \cdot T_{n,i} + (1 - \hat{c}_{r,i}) \cdot \hat{\mu}_{T_{n,i}},$$

$$\hat{\mu}_{T_{n,i}} = \hat{E}(T_{n,i})$$

$$\hat{c}_{r,i} = \frac{n \cdot \widehat{\text{Var}}(E(T_{n,i})|\theta_i)}{n \cdot \widehat{\text{Var}}(E(T_{n,i})|\theta_i) + \hat{E}(\text{Var}(T_{n,i}|\theta_i))}.$$

The estimators $\hat{E}(\text{Var}(T_{n,i}|\theta_i))$ and $\widehat{\text{Var}}(E(T_{n,i}|\theta_i))$ are explained below. They are the ones suggested by Gisler, Reinhard [Gisler, Reinhard 1993].

Let X_{ij} be the jth claim amount of risk i, $i = 1, \ldots, N$, $j = 1, \ldots, n$. Obviously, from

$$T_{n,i} = \frac{1}{n(1+a)} \sum_{j=1}^{n} \max \left\{ (1 + a - b) \cdot T_{n,i}, \min \left\{ X_{ij}, (1 + a + b) \cdot T_{n,i} \right\} \right\}$$

$$\tilde{T}_{n,i} = \frac{1}{n(1+a)d} \sum_{j=1}^{n} \max \left\{ (1 + a - b) \cdot T_{n,i}, \min \left\{ X_{ij}, (1 + a + b) \cdot T_{n,i} \right\} \right\}$$

it follows immediately that neither $T_{n,i}$ nor $\tilde{T}_{n,i}$ are described as a sum of independent random variables. Nevertheless, Gisler Reinhard treat them similarly [Gisler, Reinhard 1993], suggesting to use the empirical asymptotic variance (2.26)

$$\hat{\sigma}_{T_{n,i}}^2 = \frac{1}{n} \widehat{\text{Var}}(IF(X; F_n, T))$$

as an estimator for $\text{Var}(T_{n,i}|\theta_i)$. Thus $m := E(\text{Var}(T_{n,i}|\theta_i))$, $\tilde{m} := E(\text{Var}(\tilde{T}_{n,i}|\theta_i))$ are estimated using

$$\hat{m} = \frac{1}{N} \cdot \frac{1}{n} \sum_{i-1}^{N} \widehat{\text{Var}}(IF(X; F_n, T)) \tag{5.6}$$

$$\hat{\tilde{m}} = \frac{1}{N} \cdot \frac{1}{n} \sum_{i=1}^{N} \widehat{\text{Var}}(IF(X; F_n, \tilde{T})). \tag{5.7}$$

We know from (2.23), (4.1) and (4.2) that the asymptotic variances of $T_{n,i}$ and $\tilde{T}_{n,i}$ are

$$\text{Var}(IF(X; F, T)) = E(IF^2(X; F, T)), \qquad \text{Var}(IF(X; F, \tilde{T})) = E(IF^2(X; F, \tilde{T})).$$

According to (2.15) and Lemma 3.5, the influence function is the derivative from the right.

We therefore use as estimator

$$\widehat{\mathrm{Var}}(T_{n,i}) = \frac{\int\limits_0^\infty \psi^2\left(\frac{s}{T_{n,i}}\right) dF_n(s)}{\left(\int\limits_0^\infty \left(\frac{\partial}{\partial\theta}\right)_+ \psi\left(\frac{s}{\theta}\right)\big|_{\theta=T_{n,i}} dF_n(s)\right)^2}$$

$$= \frac{\frac{1}{n}\sum\limits_{j=1}^n \psi^2\left(\frac{X_{ij}}{T_{n,i}}\right)}{\left(\frac{1}{n}\sum\limits_{j=1}^n \frac{1}{T_{n,i}}\cdot\psi\left(\frac{X_{ij}}{T_{n,i}}\right)\right)^2}.$$

Changing the norming constant from $1/n$ to $1/(n-1)$ and noting that

$$\left(\frac{d}{dx}\right)_+ \psi(x) = \begin{cases} 1 & 1+a-b \le x \le 1+a+b \\ 0 & \text{else} \end{cases}$$

$$\left(\frac{d}{dx}\right)_+ \tilde{\psi}(x) = \begin{cases} \frac{1}{d} & 1+a-b \le x \le 1+a+b \\ 0 & \text{else} \end{cases}$$

we get

$$\widehat{\mathrm{Var}}(T_{n,i}) = \frac{1}{\left((\alpha+a)-\frac{1}{n}\sum\limits_{j=1}^n \mathbb{1}_{(0,(1+a-b)T_{n,i})\cup((1+a+b)T_{n,i},\infty)}(X_{ij})\right)^2}$$

$$\cdot\left[\frac{1}{n-1}\sum\limits_{j=1}^n \left(\max\left\{(1+a-b)\cdot T_{n,i}, \min\left\{X_{ij}, (1+a+b)\cdot T_{n,i}\right\}\right\}\right.\right.$$

$$\left.\left. - (1+a)\cdot T_{n,i}\right)^2\right]$$

and

$$\widehat{\mathrm{Var}}(\tilde{T}_{n,i}) = \frac{1}{\left((\alpha+a)d-\frac{1}{n}\sum\limits_{j=1}^n \mathbb{1}_{(0,(1+a-b)\tilde{T}_{n,i})\cup((1+a+b)\tilde{T}_{n,i},\infty)}(X_{ij})\right)^2}$$

$$\cdot\left[\frac{1}{n-1}\sum\limits_{j=1}^n \left(\max\left\{(1+a-b)\cdot \tilde{T}_{n,i}, \min\left\{X_{ij}, (1+a+b)\cdot \tilde{T}_{n,i}\right\}\right\}\right.\right.$$

$$\left.\left. - (1+a)d\cdot \tilde{T}_{n,i}\right)^2\right].$$

For a detailed calculation we refer to Appendix A.10.

Unfortunately, we do not know much about $E(T_{n,i}|\theta)$ either. But it seems reasonable to choose $T_{n,i}$ as an estimator. Then, since

$$v := \mathrm{Var}(E(T_{n,i}|\theta_i)) = \mathrm{Var}(T_{n,i}) - E(\mathrm{Var}(T_{n,i}|\theta_i)) = \mathrm{Var}(T_{n,i}) - m$$

we use as estimators

$$\hat{v} := \widehat{\mathrm{Var}}(E(T_{n,i})) = \widehat{\mathrm{Var}}(T_{n,i}) - m = \frac{1}{N-1} \sum_{i=1}^{n} \left(T_{n,i} - \bar{T}_n\right)^2 - \hat{m}$$

$$\hat{\tilde{v}} := \widehat{\mathrm{Var}}(E(\tilde{T}_{n,i})) = \widehat{\mathrm{Var}}(\tilde{T}_{n,i}) - m = \frac{1}{N-1} \sum_{i=1}^{n} \left(\tilde{T}_{n,i} - \bar{\tilde{T}}_n\right)^2 - \hat{\tilde{m}}.$$

Our empirical robust estimator of the ordinary part of the credibility premium is then

$$\hat{\mu}_{\mathrm{rC},i;n} = \hat{c}_{r,i} \cdot T_{n,i} + (1 - \hat{c}_{r,i}) \cdot \hat{\mu}_{T_{n,i}},$$
$$\hat{\mu}_{T_{n,i}} = \hat{E}(T_{n,i})$$
$$\hat{c}_{r,i} = \frac{\hat{m}}{\hat{v} + \hat{m}} \cdot T_{n,i} + \frac{\hat{v}}{\hat{v} + \hat{m}} \cdot \hat{\mu}_{T_{n,i}}$$

To get an estimator of the credibility premium $\mu_{\mathrm{ind},i}$ of risk class i, we need an estimator for μ_{xs}, too. As suggested in Section 1.4, we use the estimator of the mean excess function $e(\ell) = E(X - \ell | X > \ell)$, $i = 1, \ldots, N$. We choose $\ell = l \cdot T_{n,i}$ or $\ell = l \cdot \tilde{T}_{n,i}$, respectively, i.e. $e(l), \tilde{e}(l)$ are the mean exceedance amounts of our claim variable over an l-multiple of the estimators $T_{n,i}, \tilde{T}_{n,i}$, where $l \gg 1$ is reasonable chosen. The estimators are

$$\hat{e}(l) = \frac{\sum_{i=1}^{N} \sum_{j=1}^{n} \mathbb{1}_{(l \cdot T_{n,i}, \infty)}(X_{ij}) \cdot (X_{ij}) - l \cdot T_{n,i}}{\sum_{i=1}^{N} \sum_{j=1}^{n} \mathbb{1}_{(l \cdot T_{n,i}, \infty)}(X_{ij})}$$

$$\hat{\tilde{e}}(l) = \frac{\sum_{i=1}^{N} \sum_{j=1}^{n} \mathbb{1}_{(l \cdot \tilde{T}_{n,i}, \infty)}(X_{ij}) \cdot (X_{ij}) - l \cdot \tilde{T}_{n,i}}{\sum_{i=1}^{N} \sum_{j=1}^{n} \mathbb{1}_{(l \cdot \tilde{T}_{n,i}, \infty)}(X_{ij})}$$

leading to estimators for the robust empirical credibility premiums

$$\hat{\mu}_{\mathrm{ind},i;n} = \frac{\hat{m}}{\hat{v} + \hat{m}} \cdot T_{n,i} + \frac{\hat{v}}{\hat{v} + \hat{m}} \cdot \hat{\mu}_{T_{n,i}} + \hat{e}(l)$$

$$\hat{\tilde{\mu}}_{\mathrm{ind},i;n} = \frac{\hat{\tilde{m}}}{\hat{\tilde{v}} + \hat{\tilde{m}}} \cdot \tilde{T}_{n,i} + \frac{\hat{\tilde{v}}}{\hat{\tilde{v}} + \hat{\tilde{m}}} \cdot \hat{\mu}_{\tilde{T}_{n,i}} + \hat{\tilde{e}}(l).$$

Since we want to compare our robust credibility premiums with the original credibility premium, we state the estimator of the latter here as well.

Again X_{ij} is the claim amount of risk class i in insurance period j. According to Section 1.2.2

$$\hat{\mu}_{\mathrm{C},i;n} = \hat{c} \cdot \bar{X}_i + (1 - \hat{c}) \cdot \hat{\mu}, \quad \bar{X}_i = \frac{1}{n} \sum_{j}^{n} X_{ij}, \quad \hat{\mu} = \frac{1}{N} \sum_{i=1}^{N} \bar{X}_i$$

$$\hat{c} = \frac{n \cdot \widehat{\mathrm{Var}}(E(X|\theta_i))}{n \cdot \widehat{\mathrm{Var}}(E(X|\theta_i)) + E(\mathrm{Var}(X|\theta_i))}$$

$$\widehat{\text{Var}}(E(X|\theta_i)) = \frac{1}{N-1} \sum_{i=1}^{N} \left[\frac{1}{n} \sum_{j=1}^{n} X_{ij} - \frac{1}{N} \sum_{i=1}^{N} \bar{X}_i \right]^2$$

$$\hat{E}(\text{Var}(X|\theta_i)) = \frac{1}{N} \sum_{i=1}^{N} \left(\frac{1}{n-1} \sum_{j=1}^{n} \left(X_{ij} - \frac{1}{n} \sum_{j=1}^{n} X_{ij} \right)^2 \right).$$

For our simulation study we assume a portfolio of 25 risk classes with a claim history of $n = 30$ years. The parameters of the structure function $\Gamma(\beta, \gamma)$, i.e. the distribution function of the i.i.d. risk parameters $\theta_1, \dots, \theta_N$ are $\beta = 1$, $\gamma = 3$ and again we choose $\alpha = 1$ in (1.10).
This means that our expected claim amount for the normal claim size is

$$E(X) = E(E(X|\theta)) = E(\theta) = 3.$$

As explained in Sections 1.4 and 5.1, we apply a Pareto distribution to model the large claims. As contamination degree we choose $\varepsilon = 0.1$. Furthermore, we set $b = 1/2 \cdot \ln 2$ for both ψ and $\tilde{\psi}$, as well as $l = 100$ to estimate μ_{xs}.
Our parameters for the Pareto distribution are $\lambda = 2$ and the minimum value x_0 is chosen to be 75 times the expected value of the normal claim size amount, i.e. $x_0 = 225$.

Having a close look at the results of our simulation study in Table 5.3, the first thing we notice is, that the credibility factors in the robust models are much smaller than in the classical credibility setup. This means, the robust estimators put more weight on the class means $\mu_{T_{n,i}}$ thus accounting for the homogeneity of the sample in the ordinary part. In contrast, the credibility factor in the classical model slightly favours the individual means. This reflects the truth as well, since the whole sample - as considered in the classical model - is inhomogeneous due to the contamination.

Again the asymptotic robust estimators estimate the expected value $E(X_{\text{ord}})$ of the ordinary claim amount quite well.
Comparing all estimates to the net premium of the overall claim amount

$$(1 - \varepsilon) \cdot E(X_{\text{ord}}) + \varepsilon \cdot E(X_{\text{xs}}) = 0.9 \cdot 3 + 0.1 \cdot 450 = 47.7,$$

we realise that the classical credibility premium tends to underestimate the net premium.

In risk classes where no excess claims have been identified, i.e. $\hat{\mu}_{\text{xs}} = 0$, the robust premiums differ much from the classical credibility premium - they are all much smaller. It seems to be reasonable to apply the classical credibility premium in those cases. The meaning behind this approach is that in the absence of other models, actuarial principles ask for a reasonable consideration and weighting of all claim sizes. But if no large claim amounts are identified, the mid size claims amounts are partly neglected due to the truncation of the claims. Thus, choosing the robust credibility premiums in such cases means that the actuarial principle is violated.

i	$T_n(\mathbf{x}; a_{\mathrm{Fsh}}, b)$ $c = 0.090$			$T_n(\mathbf{x}; 0, b)$ $c = 0.087$			$\tilde{T}_n(\mathbf{x}; a_{\mathrm{MV}}, b)$ $c = 0.391$			$\tilde{T}_n(\mathbf{x}; b-1, b)$ $c = 0.083$			$\bar{X}$ $c = 0.491$
	$\hat{\mu}_{rC,i;n}$	$\hat{\mu}_{xs}$	$\hat{\mu}_{\mathrm{ind},i;n}$	$\hat{\mu}_{rC,i;n}$	$\hat{\mu}_{xs}$	$\hat{\mu}_{\mathrm{ind},i;n}$	$\hat{\mu}_{rC,i;n}$	$\hat{\mu}_{xs}$	$\hat{\mu}_{\mathrm{ind},i;n}$	$\hat{\mu}_{rC,i;n}$	$\hat{\mu}_{xs}$	$\hat{\mu}_{\mathrm{ind},i;n}$	$\hat{\mu}_{C,i;n}$
1	5.907	265.330	271.237	3.497	244.038	247.535	14.208	261.626	275.834	4.898	237.347	242.244	40.825
2	5.405	265.330	270.735	3.194	244.038	247.232	13.909	261.626	275.535	4.398	237.347	241.745	80.376
3	5.565	265.330	270.895	3.289	244.038	247.327	13.154	261.626	274.780	4.452	237.347	241.799	47.335
4	5.524	265.330	270.854	3.245	244.038	247.283	13.090	261.626	274.716	4.538	237.347	241.885	43.507
5	5.786	265.330	271.116	3.438	244.038	247.477	78.914	261.626	340.541	4.547	237.347	241.894	63.984
6	5.513	265.330	270.843	3.248	244.038	247.287	13.144	261.626	274.770	4.406	237.347	241.752	38.124
7	6.479	265.330	271.808	3.787	244.038	247.825	15.559	261.626	277.185	4.911	237.347	242.257	40.516
8	6.064	265.330	271.394	3.552	244.038	247.590	14.962	261.626	276.588	4.815	237.347	242.161	67.533
9	5.412	265.330	270.742	3.196	244.038	247.234	12.548	261.626	274.174	4.434	237.347	241.781	45.184
10	5.323	265.330	270.653	3.138	244.038	247.176	12.227	261.626	273.853	4.356	237.347	241.703	31.023
11	5.229	265.330	270.559	3.089	244.038	247.127	11.894	261.626	273.520	4.271	237.347	241.617	35.124
12	5.412	265.330	270.742	3.189	244.038	247.227	12.503	261.626	274.129	4.442	237.347	241.789	51.897
13	5.416	265.330	270.746	3.182	244.038	247.221	12.528	261.626	274.154	4.446	237.347	241.792	38.866
14	5.323	265.330	270.653	3.139	244.038	247.177	12.315	261.626	273.941	4.338	237.347	241.684	62.055
15	5.970	265.330	271.300	3.519	244.038	247.558	15.135	261.626	276.762	4.826	237.347	242.173	49.541
16	5.504	265.330	270.833	3.247	244.038	247.285	12.797	261.626	274.423	4.462	237.347	241.808	57.344
17	5.528	265.330	270.858	3.244	244.038	247.283	13.371	261.626	274.997	4.464	237.347	241.811	40.305
18	6.300	265.330	271.629	3.621	244.038	247.659	83.658	261.626	345.285	5.259	237.347	242.606	85.218
19	5.557	265.330	270.887	3.271	244.038	247.309	14.855	261.626	276.481	4.527	237.347	241.873	61.354
20	5.555	265.330	270.885	3.295	244.038	247.333	12.845	261.626	274.471	4.561	237.347	241.908	53.761
21	6.656	265.330	271.986	3.940	244.038	247.978	18.396	261.626	280.022	5.256	237.347	242.603	49.676
22	5.464	265.330	270.794	3.226	244.038	247.265	12.655	261.626	274.281	4.423	237.347	241.770	35.906
23	5.303	265.330	270.633	3.123	244.038	247.161	12.605	261.626	274.231	4.311	237.347	241.658	56.808
24	5.544	265.330	270.874	3.264	244.038	247.302	13.117	261.626	274.743	4.529	237.347	241.876	44.486
25	5.864	265.330	271.194	3.437	244.038	247.475	15.429	261.626	277.055	4.753	237.347	242.100	40.188

Table 5.3: Robust Credibility Premium $\hat{\mu}_{rC,i;n}$, Large Claim Premium $\hat{\mu}_{xs}$, Individual Premium $\hat{\mu}_{\mathrm{ind},i;n}$ and Classical Credibility Premium $\mu_{C,i;n}$ for an Insurance Portfolio with $N = 25$ Risk Classes

We also clearly see that the present choice of the estimator for μ_{xs} is not smart. Especially since the factor l can be chosen arbitrary by the actuary so far, the mean excess does not seem to be a good choice. Indeed, it is a certain mean value, thus facing the same problems as the ordinary mean estimator.

So, the actuary, pricing claims that are large in rare cases, can draw several conclusions from these results.
First of all, using a robust estimator enables him to test the portfolio regarding large claims. If this test is positive, the robust estimator gives a comparatively good prediction of the risk parameter $\theta_i, i = 1, \ldots, N$. If no large claims can be detected, the actuary should stay with the classical credibility premium. Otherwise, applying the robust credibility approach, the resulting premium calculation method will partly ignore mid-size claim amounts.

Conclusion and Outlook

In this work, we dealt with the problem of estimating an insurance premium when rare excess claims cannot be excluded. The data model for the claim amounts, we applied in this thesis reflects this by a contamination of the claim amount distribution F_θ

$$\mathcal{F}_\varepsilon(F_\theta) = \{F : F = (1 - \varepsilon) \cdot F_\theta + \varepsilon \cdot G, \ G \text{ distribution function}\}, \qquad 0 < \varepsilon < 1.$$

We chosed F_θ to be the distribution function of $\Gamma(\alpha, \theta)$ and G to be the distribution function of $Par(\lambda, x_0)$, representing common choices of claim amount distribution functions in insurance mathematics.

Therefore, our premium model consisted of an ordinary part μ_{ord} and an excess part μ_{xs}. To determine the premium for the ordinary part, we applied the idea of a credibility premium introduced by Bühlmann [Bühlmann 1967].
The central idea in credibility theory is, that in each portfolio further risk classes can be identified. And each risk class is associated with a risk parameter θ. In our case, the premium μ_{ord} depends on the risk parameter, $\mu_{\mathrm{ord}} = \mu(\theta)$.

Because of the possible large claim amounts, we substituted a robust estimator T_n for the individual mean $\bar{X}$ to get the empirical robust the credibility premium

$$\hat{\mu}_{\mathrm{C};n}(\theta) = \hat{c} \cdot T_n + (1 - \hat{c}) \cdot \mu_{T_n}.$$

Indeed, since in our estimation problem a scale parameter is engaged, we focused on robust scale M-estimators defined by

$$\sum_{j=1}^n \psi\left(\frac{X_j}{T_n}\right) = 0,$$

where $X_1, \ldots, X_n$ is the claim history of the considered risk class.

For the function ψ we gave two definitions.
Choosing ψ to be

$$\psi(x) = \max\left\{-b, \min\left\{x - (\alpha + a), b\right\}\right\}, \qquad b > 0, \ a = a(b),$$

the resulting estimator T_n was proved to be Fisher-consistent with minimal asymptotic variance.

The drawback of T_n as being difficult to calculate was compensated when instead of ψ, we applied $\bar{\psi}(x) = \frac{1}{d}(\psi(x) + (\alpha + a)) - (\alpha + a)$ to get the robust M-estimator $\tilde{T}_n$.

A main part of the thesis was devoted to the determination of optimal values for a in ψ and $\tilde{\psi}$.
According to Lemma 2.10, employing the function ψ leads to a Fisher-consistent M-estimator if

$$\int\limits_0^\infty \max\left\{-b, \min\left\{\frac{s}{\theta} - (\alpha + a), b\right\}\right\} dF_\theta(s) = 0.$$

So our optimal choice for parameter a was a_{Fish} solving the above equation.
Finding an optimal value for a in $\tilde{\psi}$, we approached the problem from the viewpoint of minimal asymptotic variance. Thus $a = a_{\mathrm{MV}}$ was chosen according to

$$a_{\mathrm{MV}} = \operatorname*{argmin}_a \operatorname{Var}(T_n(a, b)).$$

In case of the claim amount distribution to be a contaminated Exponential distribution, we were able to prove that for all $b > 0$ there exists a unique $a^\star > b - 1$ such that $a^\star = a_{\mathrm{MV}}$. Additionally, in both cases we stated the asymptotic values for a_{Fish} and a_{MV} as $b \to \infty$.

For both estimators we proved the existence of the influence functions $IF(x; F, T)$ and $IF(x; F, \tilde{T})$ which enabled us to study certain quantitative robustness characteristics. Namely, we examined the asymptotic normality, the gross error and the finite sample breakdown point. Doing so, we discovered that for the estimator T_n with a chosen to be equal to a_{Fish}, the optimal breakdown point $\varepsilon^\star = \frac{1}{2}$ cannot be gained due to the restriction $b > 0$.

To demonstrate their ability, our estimators underwent a simulation study. It turned out that the parameter b should be chosen with great care because too big values of b will yield estimates being greater than the mean value. For comparetively small values of b, the estimation works quite well. It is interesting to note that the inherent bias of both estimators T_n and $\tilde{T}_n$ vanishes when the asymptotic values $a = 0$ and $a = b - 1$ are employed.

The thesis closed with a simulation of an insurance portfolio consisting of 25 risk classes. Again the robust estimators gave a good prediction of the premium's ordinary part. Moreover, the credibility factors reflected the homogeneity of the ordinary claims by shifting weight to the class mean. But we also figured out that the choice of the mean excess function to be used as an estimator for the premium's excess part should be reconsidered.

At this stage we also want to give an outlook on topics that are interesting in connection with this work.
Certainly more research on the estimation of μ_{xs} should be done with emphasis lieing on both statistical and actuarial needs. As we discussed in Chapter 5, applying the mean

excess function for estimation μ_{xs}, the actuary is confronted with the same pitfalls as in the case of the arithmetic mean.

In the wake of Solvency II getting more important for insurance companies, the Value-at-Risk is often used in the context of extreme losses. But the Value-at-Risk might be too general in its assumptions, since it does not take into account the claim amount's distribution above a certain threshold. Nevertheless, it is in some sense an appropriate risk measure because of being a statistical tool meeting actuarial requirements. Recent extensions such as the Tail Value-at-Risk and other expected shortfall related risk measures result in a very high solvency capital, thus forcing the insurance company to keep a large amount of money in the reserve.

And just as the classical Bühlmann credibility model has been generalised to be applied for claim amount histories of different lengths, the same can be done in a robust context, cf. Gisler, Reinhard [Gisler, Reinhard 1993].

Furthermore the assumption of claim amount distributions being contaminated Exponential distributions is very limiting. Numerical evaluations suggest that the results of Chapter 3 do hold for $\alpha > 1$ as well. But we still lack a proof.

Another interest subject worth looking at, is the definition of the contamination class $\mathcal{F}_\varepsilon(F_\theta)$, cf. Section 1.4. Thall [Thall 1979] examines contamination classes described as an ε-neighbourhood of F_θ rather than a convex combination of distribution functions, as in our case. Thall develops an estimator as the solution to a minimax problem with respect to the asymptotic variance.
Even though Thall's estimator is restricted to very small contamination levels ($\varepsilon < 0.0095$), it might be interesting to compare his results to the estimators derived here. A possible problem to study is how much impact Thall's general contamination class has on the quality
of the resulting estimators – both in the sense of variance and accuracy.

Last but not least the problem of unbiasedness is out on the table. Even though choosing a to be equal to the asymptotic values $a_{\text{Fish,asym}} = 0$ or $a_{\text{MV,asym}} = b - 1$ takes away most of the bias, a theoretical foundation would make these choices more confident.

Summarising this thesis' results, we showed that employing robust estimators in determining a credibility premium enables the actuary to get detailed information on the insurance portfolio facing large claims. Not only is he able to estimate the risk parameter more precisely, he also can identify large claims.
In the aim of Solvency II, this is a useful tool because insurance companies have to quantify their solvency capital in the light of possible large, but unlikely claims like the 50, 100 and 200 year damage. Besides our results support an actuary in an insurance company facing an increase of claim amounts originating from an increasing occurence of natural desasters.

Appendix

A.1 The Distribution Function H_{t+h}, Section 2.3.3

Recall the definition of H_t,

$$H_t = (1-t)F_1 + t\Delta_x,$$

where F_1 is the Exponential distribution with parameter $\theta = 1$. Then for H_{t+h}, $h > 0$ we get

$$
\begin{aligned}
H_{t+h} &= (1-t-h)F_1 + (t+h)\Delta_x \\
&= (1-t)F_1 + t\Delta_x - hF_1 + h\Delta_x \\
&= H_t - \frac{1-t}{1-t} \cdot hF_1 + \frac{1-t}{1-t} \cdot h\Delta_x \\
&= H_t - \frac{h}{1-t} \cdot (1-t)F_1 - \frac{h}{1-t} \cdot t\Delta_x + \frac{h}{1-t} \cdot \Delta_x \\
&= \left(1 - \frac{h}{1-t}\right) \cdot H_t + \frac{h}{1-t} \cdot \Delta_x.
\end{aligned}
$$

A.2 Scale and Location Parameter, Section 3.2.1

Let X be a random variable with continuous distribution function F_θ, where θ is a scale parameter. Then for the random variable $Y = \ln X$ with distribution function G_μ we have

$$
\begin{aligned}
G_\mu(y) &= P_{Y,\mu}(Y \leq y) = P_{X,\theta}(\ln X \leq y) \\
&= P_{X,\theta}(X \leq e^y) \\
&= P_{X,1}(X \leq e^{y - \ln\theta}) \\
&= P_{X,1}(\ln X \leq y - \ln\theta) \\
&= G_{Y,\mu=\ln 1}(y - \ln\theta) = G_{Y,0}(y - \ln\theta)
\end{aligned}
$$

with $\mu = \ln\theta$. That means $\mu = \ln\theta$ is the location parameter of the distribution function of the random variable $Y = \ln X$.

A.3 The Differentiability of ρ, Section 3.2.1

Recall that

$$\rho_{\text{scale}} : \mathbb{R} \to \mathbb{R}, \qquad \rho_{\text{scale}}(x) = \begin{cases} -bx - \frac{1}{2}(\alpha + a - b)^2 & 0 \le x < \alpha + a - b \\ \frac{1}{2}x^2 - (\alpha + a)x & \alpha + a - b \le x \le \alpha + a + b \\ bx - \frac{1}{2}(\alpha + a + b)^2 & x > \alpha + a + b. \end{cases}$$

The crucial points for the differentiablity of ρ are $x = \alpha + a - b$ and $x = \alpha + a + b$. But it follows immediately that

$$\rho'(\alpha + a - b) = -b = \lim_{x \to (\alpha + a - b)-} \rho'(x)$$

and

$$\rho'(\alpha + a + b) = b = \lim_{x \to (\alpha + a + b)+} \rho'(x)$$

Thus ρ is differentiable.

A.4 Existence of the Influence Function, Section 3.2.2

Denote by H_t the convex combination of $F_1 = \Gamma(\alpha, 1)$ and Δ_x, cf. Appendix A.1. Then

$$\nabla := \frac{\int_0^\infty \psi\left(\frac{s}{T(H_t)}\right) dH_t(s) - \int_0^\infty \psi\left(\frac{s}{T(H_0)}\right) dH_0(s)}{t}$$

$$= \frac{(1-t)\int_0^\infty \psi\left(\frac{s}{T(H_t)}\right) dF(s) + t \cdot \psi\left(\frac{x}{T(H_t)}\right) - \int_0^\infty \psi\left(\frac{s}{T(H_0)}\right) dH_0(s)}{t}$$

$$\stackrel{(\star)}{=} \frac{(1-t)\left(\int_0^\infty \psi\left(\frac{s}{T(H_t)}\right) dF(s) - \int_0^\infty \psi\left(\frac{s}{T(H_0)}\right) dH_0(s)\right) - t \cdot \int_0^\infty \psi\left(\frac{s}{T(H_0)}\right) dH_0(s)}{t}$$

$$\quad + \psi\left(\frac{x}{T(H_t)}\right)$$

$$= (1-t) \cdot \underbrace{\frac{\int_0^\infty \psi\left(\frac{s}{T(H_t)}\right) dF(s) - \int_0^\infty \psi\left(\frac{s}{T(H_0)}\right) dH_0(s)}{t}}_{(I)} - \underbrace{\int_0^\infty \psi\left(\frac{s}{T(H_0)}\right) dH_0(s)}_{(II)}$$

$$\quad + \psi\left(\frac{x}{T(H_t)}\right),$$

where $(\star)$ is due to the product rule of differentiation.

Because of the assumptions of Lemma 3.5, we know that for term (II)

$$\int\limits_0^\infty \psi\left(\frac{s}{T(H_0)}\right) dH_0(s) = \int\limits_0^\infty \psi\left(\frac{s}{T(F_1)}\right) dF_1(s) = 0.$$

Recalling the definition of the function ψ, namely

$$\psi(x) = \max\left\{-b, \min\left\{x - (\alpha + a), b\right\}\right\},$$

we get for term (I)

$$
\begin{aligned}
(I) = \frac{1}{t} \cdot \Bigg[& -b \cdot F_1\big((\alpha + a - b)T(H_t)\big) + \int\limits_{(\alpha + a - b)T(H_t)}^{(\alpha + a + b)T(H_t)} \frac{s}{T(H_t)} dF_1(s) \\
& - (\alpha + a) \cdot F_1\big((\alpha + a + b)T(H_t)\big) + (\alpha + a) \cdot F_1\big((\alpha + a - b)T(H_t)\big) \\
& + b \cdot \Big(1 - F_1\big((\alpha + a + b)T(H_t)\big)\Big) \\
& - \Bigg[-b \cdot F_1\big((\alpha + a - b)T(H_0)\big) + \int\limits_{(\alpha + a - b)T(H_0)}^{(\alpha + a + b)T(H_0)} \frac{s}{T(H_0)} dF_1(s) \\
& - (\alpha + a) \cdot F_1\big((\alpha + a + b)T(H_0)\big) + (\alpha + a) \cdot F_1\big((\alpha + a - b)T(H_0)\big) \\
& + b \cdot \Big(1 - F_1\big((\alpha + a + b)T(H_0)\big)\Big) \Bigg] \Bigg] \\
= \frac{1}{t} \cdot \Bigg[& \underbrace{(\alpha + a - b)\Big[F_1\big((\alpha + a - b)T(H_t)\big) - F_1\big((\alpha + a - b)T(H_0)\big)\Big]}_{(Ia)} \\
& + \underbrace{(\alpha + a + b)\Big[1 - F_1\big((\alpha + a + b)T(H_t)\big) - 1 + F_1\big((\alpha + a + b)T(H_0)\big)\Big]}_{(Ib)} \\
& - (\alpha + a) + (\alpha + a) \\
& + \underbrace{\int\limits_{(\alpha + a - b)T(H_t)}^{(\alpha + a + b)T(H_t)} \frac{s}{T(H_t)} dF_1(s) - \int\limits_{(\alpha + a - b)T(H_0)}^{(\alpha + a + b)T(H_0)} \frac{s}{T(II_0)} dF_1(s)}_{(Ic)} \Bigg].
\end{aligned}
$$

Again we analyse the different expressions separately. For (Ia) it is easy to see

$$
\frac{F_1\big((\alpha + a - b)T(H_t)\big) - F_1\big((\alpha + a - b)T(H_0)\big)}{t}
$$

$$
= \frac{F_1\big((\alpha + a - b)T(H_t)\big) - F_1\big((\alpha + a - b)T(H_0)\big)}{(\alpha + a - b)\Big[T(H_t) - T(H_0)\Big]} \cdot \frac{(\alpha + a - b)\Big[T(H_t) - T(H_0)\Big]}{t}.
$$

In the same manner for term (Ib) we have

$$-\frac{F_1\big((\alpha+a+b)T(H_t)\big)-F_1\big((\alpha+a+b)T(H_0)\big)}{t}$$

$$=-\frac{F_1\big((\alpha+a+b)T(H_t)\big)-F_1\big((\alpha+a+b)T(H_0)\big)}{(\alpha+a+b)\Big[T(H_t)-T(H_0)\Big]}\cdot\frac{(\alpha+a+b)\Big[T(H_t)-T(H_0)\Big]}{t}.$$

For the third expression (Ic) it needs a bit more calculation

$$\frac{\displaystyle\int_{(\alpha+a-b)T(H_t)}^{(\alpha+a+b)T(H_t)}\frac{s}{T(H_t)}\,dF_1(s)-\int_{(\alpha+a-b)T(H_0)}^{(\alpha+a+b)T(H_0)}\frac{s}{T(H_0)}\,dF_1(s)}{t}$$

$$=\frac{\displaystyle\int_{(\alpha+a-b)T(H_t)}^{(\alpha+a+b)T(H_t)}\frac{s}{T(H_t)}\,dF_1(s)-\int_{(\alpha+a-b)T(H_0)}^{(\alpha+a+b)T(H_0)}\frac{s}{T(H_t)}\,dF_1(s)}{t}$$

$$+\frac{\displaystyle\int_{(\alpha+a-b)T(H_0)}^{(\alpha+a+b)T(H_0)}\left(\frac{s}{T(H_t)}-\frac{s}{T(H_0)}\right)\,dF_1(s)}{t}$$

$$=\frac{\displaystyle\int_{0}^{(\alpha+a+b)T(H_t)}\frac{s}{T(H_t)}\,dF_1(s)-\int_{0}^{(\alpha+a+b)T(H_0)}\frac{s}{T(H_t)}\,dF_1(s)}{(\alpha+a+b)\Big[T(H_t)-T(H_0)\Big]}\cdot\frac{(\alpha+a+b)\Big[T(H_t)-T(H_0)\Big]}{t}$$

$$-\frac{\displaystyle\int_{0}^{(\alpha+a-b)T(H_t)}\frac{s}{T(H_t)}\,dF_1(s)-\int_{0}^{(\alpha+a-b)T(H_0)}\frac{s}{T(H_t)}\,dF_1(s)}{(\alpha+a-b)\Big[T(H_t)-T(H_0)\Big]}\cdot\frac{(\alpha+a-b)\Big[T(H_t)-T(H_0)\Big]}{t}$$

$$-\frac{T(H_t)-T(H_0)}{t}\cdot\frac{1}{T(H_t)\cdot T(H_0)}\cdot\int_{(\alpha+a-b)T(H_0)}^{(\alpha+a+b)T(H_0)}s\,dF_1(s).$$

Recall that F_1 is a continuous distribution function. Denote by f_1 its density function. Thus letting $t\to 0$, we get for (Ia)

$$(Ia)\xrightarrow{t\to 0}(\alpha+a-b)^2\cdot f_1\big((\alpha+a-b)T(H_0)\big)$$

and for (Ib)

$$(Ib)\xrightarrow{t\to 0}-(\alpha+a+b)^2\cdot f_1\big((\alpha+a+b)T(H_0)\big).$$

Expression (Ic) converges for $t\to 0$

$$(Ic)\xrightarrow{t\to 0}(\alpha+a+b)^2\cdot f_1\big((\alpha+a+b)T(H_0)\big)-(\alpha+a-b)^2\cdot f_1\big((\alpha+a-b)T(H_0)\big)$$

$$-\frac{1}{T^2(H_0)}\cdot\int_{(\alpha+a-b)T(H_0)}^{(\alpha+a+b)T(H_0)}s\,dF_1(s)\cdot\lim_{t\to 0}\frac{T(H_t)-T(H_0)}{t}.$$

For our original expression $\bigtriangledown$ this means

$$\bigtriangledown \xrightarrow{t \to 0} -\frac{1}{T^2(H_0)} \cdot \int\limits_{(\alpha + a - b)T(H_0)}^{(\alpha + a + b)T(H_0)} s \, dF_1(s) \cdot \lim_{t \to 0} \frac{T(H_t) - T(H_0)}{t} + \psi \left(\frac{x}{T(H_0)} \right).$$

Because T is the defining funtional of an M-estimator, it also clear that $\bigtriangledown = 0$, i.e. the above limit is equal to zero. Due to $\alpha + a + b > 0$ the integral is positive. It follows

$$\lim_{t \to 0} \frac{T(H_t) - T(H_0)}{t} = \lim_{t \to 0} \frac{T((1-t)F + t\Delta_x) - T(F)}{t}$$

exists and because of the definition of the influence function (2.15)

$$IF(x; F, T) = \frac{1}{\int\limits_{(\alpha + a - b)T(H_0)}^{(\alpha + a + b)T(H_0)} s \, dF_1(s)} \cdot \psi \left(\frac{x}{T(H_0)} \right) \cdot T^2(H_0)$$

A.5 The Function h_2, Section 3.3.1

It is shown that $h_2(b-1) = be^{b-1} > e^{-(1-b)} - e^{-(1+b)}$ for $b > 1$.
Define $h_3(b) := be^{b-1} - e^{-(1-b)} + e^{-(1+b)}$, then $h_3(0) = -e^{-1} + e^{-1} = 0$. Differentiating h_3 with respect to b gives

$$h_3'(b) = e^{b-1} + be^{b-1} - e^{-(1-b)} - e^{-(1+b)}$$

$$= e^{-1} \left(be^b - e^{-b} \right)$$

$$= e^{-1} \sum_{n=0}^{\infty} \left(\frac{b^{n+1}}{n!} - \frac{(-b)^n}{n!} \right)$$

$$> 0$$

for all $b > 1$.

A.6 Expected Value of $\psi(X)$, Section 3.3.2

We have $F_1 = \Gamma(\alpha, 1)$

$$E(\psi(X)) = \int\limits_0^\infty \psi(s) \, dF_1(s)$$

$$= -bF_1(\alpha + a - b) + b(1 - F_1(\alpha + a + b)) + \int\limits_{\alpha + a - b}^{\alpha + a + b} (s - (\alpha + a)) \, dF_1(s)$$

$$= (\alpha + a - b)F_1(\alpha + a - b) + (\alpha + a + b)(1 - F_1(\alpha + a + b))$$

$$+ \int\limits_{\alpha + a - b}^{\alpha + a + b} s \, dF_1(s) - (\alpha + a)$$

$$= (\alpha + a)d - (\alpha + a)$$

with the apparent definition

$$(\alpha+a)d = (\alpha+a-b)F_1(\alpha+a-b) + (\alpha+a+b)(1-F_1(\alpha+a+b)) + \int\limits_{\alpha+a-b}^{\alpha+a+b} s\,dF_1(s)$$

A.7 Variance of $IF(X; F_1, \tilde{T})$, Section 3.3.2

$$E\left(IF^2(X; F_1, \tilde{T})\right) = \frac{E\left(\tilde{\psi}^2(X)\right)}{\frac{1}{d}\left(\int\limits_{\alpha+a-b}^{\alpha+a+b} s\,dF_1(s)\right)^2}$$

$$= \frac{E\left(\psi(X)+(\alpha+a)-(\alpha+a)d\right)^2}{\left(\int\limits_{\alpha+a-b}^{\alpha+a+b} s\,dF_1(s)\right)^2}$$

$$= \frac{E\left(\psi(X)+(\alpha+a)\right)^2 - 2(\alpha+a)d \cdot E\left(\psi(X)+(\alpha+a)\right) + (\alpha+a)^2 d^2}{\left(\int\limits_{\alpha+a-b}^{\alpha+a+b} s\,dF_1(s)\right)^2}$$

$$= \frac{E\left(\psi^2(X)\right) + 2(\alpha+a)\cdot E\left(\psi(X)\right) + (\alpha+a)^2 - (\alpha+a)^2 d^2}{\left(\int\limits_{\alpha+a-b}^{\alpha+a+b} s\,dF_1(s)\right)^2}$$

$$= \frac{E\left(\psi^2(X)\right) - ((\alpha+a)d - (\alpha+a))^2}{\left(\int\limits_{\alpha+a-b}^{\alpha+a+b} s\,dF_1(s)\right)^2}$$

$$= \frac{Var\left(\psi(X)\right)}{\left(\int\limits_{\alpha+a-b}^{\alpha+a+b} s\,dF_1(s)\right)^2}.$$

A.8 Behaviour of $\sinh(x)$, Section 3.3.2

Recall that $\sinh(x) = \frac{e^x - e^{-x}}{2}$. We want to show that $x < \sinh(x)$ for $x > 0$.

First note that $\sinh(0) = 0$. Then consider

$$\sinh'(x) = \cosh(x) = \frac{e^x + e^{-x}}{2} = \sum_{k=0}^{\infty} \frac{x^{2k}}{(2k)!} = 1 + \underbrace{\sum_{k=1}^{\infty} \frac{x^{2k}}{(2k)!}}_{>0} > 1.$$

Because $\frac{d}{dx}\,x = 1$ it follwos immediately that $x < \sinh(x)$ for $x > 0$.

A.9 Partial Derivative of h with respect to b, Section 3.3.2

Recall
$$\lim_{a \to (b-1)+} h(a,b) = \left(e^{2b} - 1\right)^3 + e^{2b}\left(e^{2b} - (1+2b)\right)\left(e^{2b} - (1+2b) - 2(e^{2b} - 1)\right).$$

Now simple calculation shows
$$\left(e^{2b} - 1\right)^3 + e^{2b}\left(e^{2b} - (1+2b)\right)\left(e^{2b} - (1+2b) - 2(e^{2b} - 1)\right)$$
$$= \left(e^{2b} - 1\right)^3 + \left(e^{4b} - e^{2b}(1+2b)\right)\left(-e^{2b} + 1 - 2b\right)$$
$$= \left(e^{2b} - 1\right)^3 + \left(e^{6b} + 2e^{4b} + 4b^2 e^{2b} - e^{2b}\right)$$
$$= -e^{4b} + 2e^{2b} + 4b^2 e^{2b} - 1$$
$$= -\left(e^{2b} - 1\right)^2 + 4b^2 e^{2b}.$$

A.10 Estimators for the Asymptotic Variance, Section 5.3

We only present the calculation for the asymptotic variance estimator of T_n. For $\tilde{T}_n$ the calculation works identically. We omit the index i for the risk class.

$$\widehat{\mathrm{Var}}(T_n) = \frac{\int_0^\infty \psi^2\left(\frac{s}{T_n}\right) dF_n(s)}{\left(\int_0^\infty \left(\frac{\partial}{\partial\theta}\right)_+ \psi\left(\frac{s}{\theta}\right)\big|_{\theta = T_n} dF_n(s)\right)^2}$$

$$= \frac{\frac{1}{n}\sum_{j=1}^n \psi^2\left(\frac{X_j}{T_n}\right)}{\left(\frac{1}{n}\sum_{j=1}^n \frac{1}{T_n} \cdot \psi\left(\frac{X_j}{T_n}\right)\right)^2}.$$

$$= \frac{\frac{1}{n}\sum_{j=1}^n \psi^2\left(\frac{X_j}{T_n}\right)}{\frac{1}{T_n^2}\left(\frac{1}{n}\sum_{j=1}^n \mathbb{1}_{[(1+a-b)T_n,(1+a+b)T_n]}(X_j) \cdot X_j\right)^2}.$$

Now, we recall the definitions of the indices l_1, l_2 in (3.5), that is
$$X_{l_1} < (1 + a - b)T_n, \qquad X_{l_1+1} \geq (1 + a - b)T_n$$
$$X_{n-l_2} < (1 + a + b)T_n, \qquad X_{l_2} > (1 + a + b)T_n$$
and formula (3.6). Then the above denominator equals

$$\frac{1}{n}\sum_{j=1}^n \mathbb{1}_{(0,(1+a-b)T_n)\cup((1+a+b)T_n,\infty)}(X_j) \cdot X_j$$

$$= T_n \cdot \left((1 + a) - \frac{1}{n}\sum_{j=1}^n \mathbb{1}_{(0,(1+a-b)T_n)\cup((1+a+b)T_n,\infty)}(X_j)\right)$$

and thus together with the fact that

$$\psi\left(\frac{x}{t}\right) = \max\left\{-b, \min\left\{\frac{x}{t} - (1+a), b\right\}\right\}$$
$$= \frac{1}{t} \cdot \max\left\{(1+a-b) \cdot t, \min\{x, (1+a+b) \cdot t\}\right\} - (1+a)$$

and a changing of the norming constant from $\frac{1}{n}$ to $\frac{1}{n-1}$ we get

$$\widehat{\mathrm{Var}}(T_n) = \frac{1}{\left((1+a) - \frac{1}{n}\sum\limits_{j=1}^{n} \mathbb{1}_{\{X \notin [(1+a-b)T_n, (1+a+b)T_n]\}}(X_j)\right)^2}$$
$$\cdot \left[\frac{1}{n-1}\sum\limits_{j=1}^{n}\left(\max\{(1+a-b)T_n, \min\{X_j, (1+a+b)T_n\}\}\right.\right.$$
$$\left.\left. - (1+a)T_n^2\right)^2\right].$$

Bibliography

[Ahrens, Dieter 1982] Ahrens, Dieter (1982), *Generating Gamma Variates by a Modified Rejection Technique*, Communications of the ACM, Vol. 25, No. 1, pp. 47-54

[Andrews et al. 1972] Andrews, Bickel, Hampel, Huber, Rogers, Tukey (1972), *Robust Estimation of Location*, Princeton University Press, Princeton

[Asmussen, Klüppelberg 1996] Asmussen, Klüppelberg (1996), *Large Deviations Results for Subexponential Tails with Applications for Insurance Risk*, Stochastic Processes and their Applications, Vol. 64, pp. 103-125

[Bain, Engelhardt 1992] Bain, Engelhardt (1992), *Introduction to Probability and Mathematical Statistics*, Duxbury Press, Belmont

[Basel 2006] Basler Ausschuss für Bankenaufsicht (2006), *Grundsätze für eine wirksame Bankenaufsicht*, www.bis.org/publ/bcbs129ger.pdf

[Bauer 1992] Bauer (1992), *Maß- und Integrationstheorie*, 2. Auflage, de Gruyter , Berlin

[Bauer 2001] Bauer (2001), *Wahrscheinlichkeitstheorie*, 5. Auflage, de Gruyter, Berlin

[Berger 1985] Berger (1985), *Statistical Decision Theory and Bayesian Analysis*, Springer, New York, 2nd Edition

[Bernardo, Smith 2000] Bernardo, Smith (2000), *Bayesian Theory*, John Wiley, Chichester

[Borch 1960] Borch (1960), *An Attempt to Determine the Optimum Amount of Stop Loss Reinsurance*, Transactions of the XVIth International Congress of Actuaries, Vol. 2, pp. 597-610

[Borch 1962] Borch (1962), *A Contribution to the Theory of Reinsurance Markets*, Skandinavisk Aktuarietidskrift, pp. 176-189

[Borch 1974] Borch (1974), *Mathematical Models in Insurance*, ASTIN Bulletin, Vol. 7, No. 3, pp. 192-202

[Bowers et al. 1997] Bowers, Gerber, Hickmann, Jones, Nesbitt (1997), *Actuarial Mathematics*, The Society of Actuaries, Schaumburg, Illinois

[Breuer, Zwas 1993] Breuer, Zwas (1993), *Numerical Mathematics*, Cambridge University Press, New York

[Browne 1995] Browne (1995), *Optimal Investment Policies for a Firm with a Random Risk Process; Exponential Utility and Minimizing the Probability of Ruin*, Mathematics of Operations Research, Vol. 20, No. 4, pp. 937-958

[Bühlmann 1967] Bühlmann (1967), *Experience Rating and Credibility*, ASTIN Bulletin, Vol. 4, pp. 119-207

[Bühlmann 1970] Bühlmann (1970), *Mathematical Methods in Risk Theory*, Springer, Berlin

[Bühlmann 1976] Bühlmann (1976), *Minimax Credibility*, Scandinavian Actuarial Journal, pp. 65-78

[Bühlmann 1980] Bühlmann (1980), *An Economic Premium Principle*, ASTIN Bulletin, Vol. 11, pp. 52-60

[Bühlmann, Gisler 2005] Bühlmann, Gisler (2005), *A Course in Credibility Theory and Its Applications*, Springer, Berlin

[Bühlmann, Straub 1970] Bühlmann, Straub (1970), *Glaubwürdigkeit für Schadensätze*, Mitteilungen der Vereinigung Schweizerischer Versicherungsmathematiker, pp. 111-133

[Bühlmann, Jewell 1979] Bühlmann, Jewell (1979), *Optimal Risk Exchanges*, ASTIN Bulletin, Vol. 10, pp. 243-262

[Chateauneuf 1999] Chateauneuf (1999), *Comonotonicity and Rank-dependent Expected Utility Theory for Arbitrary Consequences*, Journal of Mathematical Economics, Vol. 32, pp. 21-45

[Christensen, Schmidli 2000] Christensen, Schmidli (2000), *Pricing Catastrophe Insurance Products Based on Actually Reported Claims*, Insurance: Mathematics and Economics, Vol. 27, pp. 189-200

[Collins 1999] Collins (1999), *Robust M-Estimators of Scale: Minimax Bias versus Maximal Variance*, Canadian Journal of Statistics, Vol. 27, No. 1, pp. 81-96

[Collins 2003] Collins (2003), *Bias-Robust L-Estimators of a Scale Parameter*, Statistics, Vol. 37, No. 4, pp. 287-304

[Cox et al. 1979] Cox, Ross, Rubinstein (1979), *Option Pricing: A Simplified Approach*, Journal of Financial Economics, Vol. 7, pp. 229-263

[Cummins, Lewis 2002] Cummins, Lewis (2002), *Catastrophic Events, Parameter Uncertainty and the Breakdown of Implicit Long-term Contracting in the Insurance Market: The Case of Terrorism Insurance*, Wharton School Center for Financial Institutions, University of Pennsylvenia, Center for Financial Institutions Working Papers, 02-40

[Dannenburg 1996] Dannenburg (1996), *Bühlmann's Credibility Premium in the Bühlmann-Straub Model*, Mitteilungen der Schweizer Aktuarvereinigung, Vol. 1, pp. 63-78

[Dassios, Jang 2003] Dassios, Jang (2003), *Pricing Catastrophe Reinsurance and Derivatives Using the Cox Process with Shot Noise Intensity*, Finance and Stochastics, Vol. 7, pp. 73-95

[Dean 2005] Dean (2005), *Topics in Credibility Theory*, Study Note C-24-05, Education and Examination Committee of the Society of Actuaries, www.soa.org/files/pdf/c-24-05.pdf

[Denuit et al. 2001] Denuit, Dhaene, Ribas (2001), *Does Positive Dependence between Individual Risks Increase Stop-Loss Premiums?*, Insurance: Mathematics and Economics, Vol. 28, No. 3, pp. 305-308

[Donoho, Huber 1983] Donoho, Huber (1983), *The Notion of Breakdown Point*, in: Bickel, Doksum, Hodges (eds.) *A Festschrift for Erich Lehmann*, Wadsworth International Group, Belmont, pp. 157-184

[Dubey, Gisler 1981] Dubey, Gisler (1981), *On Parameter Estimators in Credibility*, Mitteilungen der Vereinigung Schweizerischer Versicherungsmathematiker, Vol. 81, No. 2, pp. 187-212

[d'Ursel, Lauwers 1985] d'Ursel, Lauwers (1985), *Chains of Reinsurance: Non-cooperative Equilibria and Pareto Optimality*, Insurance: Mathematics and Economics, Vol. 4, pp. 279-285

[Eichenauer et al. 1988] Eichenauer, Lehn, Rettig (1988), *A Gamma-minimax Result in Credibility Theory*, Insurance: Mathematics and Economics, Vol. 1, pp. 49-57

[Embrechts, Schmidli 1994] Embrechts, Schmidli (1994), *Modelling of Extremal Events in Insurance and Finance*, Mathematical Methods of Operations Research, Vol. 39, pp. 1-34

[Embrechts et al. 2003] Embrechts, Klüppelberg, Mikosch (2003), *Modelling Extremal Events*, Springer, Berlin

[Extremus] Extremus Versicherungs-AG, www.extremus-online.de

[Fisz 1966] Fisz (1966), *Wahrscheinlichkeitsrechnung und Mathematische Statistik*, Deutscher Verlag der Wissenschaften, Berlin

[Froot 2001] Froot (2001), *The Market for Catastrophic Risk: A Clinical Examination*, Journal of Financial Economics, Vol. 60, pp. 529-571

[Gather, Schultze 1999] Gather, Schultze (1999), *Robust Estimation of Scale of an Exponential Distribution*, Statistica Neerlandica, Vol. 53, No. 3, pp. 327-341

[Gerber 1974] Gerber (1974), *On Additive Premium Calculation Principles*, ASTIN Bulletin, Vol. 7, No. 3, pp. 215-222

[Gerber 1977] Gerber (1977), *On the Computation of Stop-Loss Premiums*, Mitteilungen der Vereinigung Schweizerischer Versicherungsmathematiker, Vol. 77, pp. 45-58

[Gerber 1979] Gerber (1979), *An Introduction to Mathematical Risk Theory*, University of Pennsylvania, Philadelphia, Pa.

[Gisler, Reinhard 1993] Gisler, Reinhard (1993), *Robust Credibility*, ASTIN Bulletin, pp. 117-143

[Goovaerts et al. 1984] Goovaerts, De Vylder, Haezendonck (1984), *Insurance Premiums*, North-Holland, Amsterdam

[Goulet 2001] Goulet (2001), *A Generalized Crossed Classification Credibility Model*, Insurance: Mathematics and Economics, Vol. 28, pp. 205-216

[Härdle, Gasser 1984] Härdle, Gasser (1984), *Robust Non-parametric Function Fitting*, Journal of the Royal Statistical Society, Series B, Vol. 46, pp. 42 - 51

[Hampel 1971] Hampel (1971), *A General Qualitative Definition of Robustness*, Annals of Mathematical Statistics, Vol. 42, pp. 1887-1896

[Hampel 1974] Hampel (1974), *The Influence Curve and Its Role in Robust Estimation*, Journal of the American Statistical Association, Vol. 69, pp. 383 -393

[Hampel et al. 1986] Hampel, Ronchetti, Rousseeuw, Stahel (1986), *Robust Statistics - The Approach Based on Influence Functions*, John Wiley, New York

[Herlihy, Parisi 1999] Herlihy, Parisi (1999), *Modelling Catastrophe Reinsurance Risk: Implications for the CAT Bond Market*, Standard & Poor's Structured Finance Special Report, New York

[Herzog 1999] Herzog (1999), *Introduction to Credibility Theory*, ACTEX Publications, Winsted, 3rd Edition

[Hogg 1979] Hogg (1979), *An Introduction to Robust Estimation*, in: Launer, Wilkinson (eds.), *Robustness in Statistics*, Academic Press, New York, pp. 1-17

[Huber 1964] Huber (1964), *Robust Estimation of a Location Parameter*, Annals of Mathematical Statistics, Vol. 35, pp. 73-101

[Huber 1981] Huber (1981), *Robust Statistics*, John Wiley, New York

[Jaffee, Russel 1996] Jaffee, Russell (1996), *Catastrophe Insurance, Capital Markets and Uninsurable Risks*, Wharton School Center for Financial Institutions, University of Pennsylvenia, Center for Financial Institutions Working Paper 96-12, http://ideas.repec.org/p/wop/pennin/96-12.html

[Jewell 1974] Jewell (1974), *Credible Means are Exact Bayesian for Exponential Families*, ASTIN Bulletin, Vol. 8, pp. 77-90

[Jewell 1976] Jewell (1976), *A Survey of Credibility Theory*, Berkely Research Center Technical Report, 1976

[Johnson et al. 1994] Johnson, Kotz, Balakrishnan (1994), *Continuous Univariate Distributions, Volume 1*, John Wiley, Toronto, 2nd Edition

[Jurečková, Picek 2006] Jurečková, Picek (2006), *Robust Statistical Methods with R*, Chapman & Hall/CRC, Boca Raton

[Kimber 1983] Kimber (1983), *Comparison of Some Robust Estimators of Scale in Gamma Samples with Known Shape*, Journal of Statistical Computational Simulation, Vol. 18, pp. 273-286

[Künsch 1992] Künsch (1992), *Robust Methods for Credibility*, ASTIN Bulletin, Vol. 22, No. 1, pp. 33-49

[Ladoucette, Teugels 2006] Ladoucette, Teugels (2006), *Analysis of Risk Measures for Reinsurance Layers*, Insurance: Mathematics and Economics, Vol. 38, pp. 630-639

[Lehmann, Casella 1998] Lehmann, Casella (1998), *Theory of Point Estimation*, Springer, New York

[Lehn, Wegmann 2004] Lehn, Wegmann (2004), *Einführung in die Statistik*, B.G. Teubner, Stuttgart, 4th Edition

[Luan 2001] Luan (2001), *Insurance Premium Calculation Principles with Anticipated Utility Theory*, ASTIN Bulletin, Vol. 31, No. 1, pp. 23-35

[Mack 1997] Mack (1997), *Schadenversicherungsmathematik*, Schriftenreihe Angewandte Versicherungsmathematik, Versicherungswirtschaft, Karlsruhe

[Mangoldt, Knopp 1975] Mangoldt, Knopp (1975), *Einführung in die Höhere Mathematik, 3. Band*, S. Hirzel, Leipzig, 14th Edition

[Marazzi 1976] Marazzi (1976), *Minimax Credibility*, Mitteilungen der Vereinigung Schweizerischer Versicherungsmathematiker, 1976, Vol. 2, pp. 219 - 229

[Martin et al. 2006] Martin, Reitz, Wehn (2006), *Kreditderivate und Kreditrisikomodelle*, Vieweg, Braunschweig/Wiesbaden

[Merz 2004] Merz (2004), *Das Konzept der orthogonalen Projektion zur Bestimmung von Credibility-Schätzern in diskreter und kontinuierlicher Zeit*, Peter Lang, Frankfurt am Main

[Mikosch 2004] Mikosch (2004), *Non-Life Insurance Mathematics*, Springer, Heidelberg

[Müller 1991] Müller (ed.) (1991), *Lexikon der Stochastik*, Akademie, Berlin

[Munich Re 2001] Munich Re ART Solutions (2001), *Risikotransfer in den Kapitalmarkt - Nutzung der Kapitalmärkte für das Management von Versicherungsrisiken*, Münchner Rückversicherungsgesellschaft, München

[Pan et al. 2008] Pan, Wang, Wu (2008), *On the Consistency of Credibility Premiums Regarding Esscher Principle*, Insurance: Mathematics and Economics, Vol. 42, No. 1, pp. 119-126

[Purcaru, Denuit 2002] Purcaru, Denuit (2002), *On the Stochastic Increasingness of Future Claims in the Bühlmann Linear Credibility Premium*, Blätter der Deutschen Gesellschaft für Versicherungsmathematik, Vol. 25, No. 4, pp. 781-793

[Resnick 1987] Resnick (1987), *Extreme Values, Regular Variation and Point Processes*, Springer, New York

[Rockafellar 1972] Rockafellar (1972), *Convex Analysis*, Princeton University Press, Princeton

[Rolski et al. 1999] Rolski, Schmidli, Schmidt, Teugels (1999), *Stochastic Processes for Insurance and Finance*. John Wiley, West Sussex

[Rousseeuw, Leroy 1987] Rousseeuw, Leroy (1987), *Robust Regression and Outlier Detection*, John Wiley, New York

[Rousseeuw, Leroy 1988] Rousseeuw, Leroy (1988), *A Robust Scale Estimator Based on the Shortest Half*, Statistica Neerlandica, Vol. 42, No. 2, pp. 103-116

[Rudin 1991] Rudin(1991), *Functional Analysis*, McGraw-Hill, Boston, 2nd Edition

[Sandström 2007] Sandström (2007), *Solvency II: Calibration for Skewness*, Scandinavian Actuarial Journal, Vol. 2, pp. 126-134

[Shiryaev 1984] Shiryaev (1984), *Probability*, Springer, New York, 2nd Edition

[Schmidli 2001] Schmidli (2001), *Optimal Proportional Reinsurance Policies in a Dynamic Setting*, Scandinavian Actuarial Journal, Vol. 1, pp. 55-68

[Schmidt 1990] Schmidt (1990), *Convergence of Bayes and Credibility Premiums*, ASTIN Bulletin, Vol. 20, No. 2, pp. 167-172

[Schmidt 2006] Schmidt (2006), *Versicherungsmathematik*, Springer, Berlin

[Sibbett 2004] Sibbett (2004), *History of Insurance*, in: Teugels (ed.) *Encyclopedia of Actuarial Science*, John Wiley, Chichester, pp. 848-862

[sigma 2003] sigma (2/2003), *Natural Catastrophes and man-made disasters in 2002: High Flood Loss Burden*, www.swissre.com

[Solvency] www.ceiops.eu

[Sundt 1999] Sundt (1999), *An Introduction to Non-Life Insurance Mathematics*, Versicherungswirtschaft, Karlsruhe

[Szatmari, Collins 2007] Szatmari-Voicu, Collins (2007), *M-Estimators of Scale with Minimum Gross Errors Sensitivity*, Communications in Statistics: Theory and Methods, Vol. 36, No. 11, pp. 2037-2048

[Thall 1979] Thall (1979), *Huber-Sense Robust M-estimation of a Scale Parameter, with Application to the Exponential Distribution*, Journal of the American Statistical Association, Vol. 74, pp. 147-152

[Tukey 1970] Tukey (1970), *Exploratory Data Analysis*, Limited Preliminary Edition, Addison-Wesley, Massachusetts

[Vesa et al. 2007] Vesa, Lasse, Raoul (2007), *Topical Modelling Issues in Solvency II*, Scandinavian Actuarial Journal, 2, pp. 135-146

[Vidakovic 2000] Vidakovic (2000), *Γ-Minimax: A Paradigm for Conservative Robust Bayesians*, in: Rios, Ruggeri (eds.) *Robust Bayesian Analysis*, Springer, New York

[Wald 1971] Wald (1971), *Statistical Decision Functions*, Chelsea, New York

[Whitney 1918] Whitney (1918), *The Theory of Experience Rating*, Proceedings of the Casualty Actuarial Society, No. 4, pp. 274-292

Wissenschaftlicher Werdegang

Persönliche Daten

Name: Annett Keller
Geburtsdatum: 25.04.1978
Geburtsort: Dresden

Schule und Studium

10/1999-12/2002   Studium an der TU Dresden, Wirtschaftsmathematik
                   Abschluß: Diplom Mathematik (Wirtschaftsmathematik)
09/1998-07/1999   Studium an der University of Calgary, Kanada, Statistik
10/1996-07/1998   Studium an der TU Dresden, Wirtschaftsmathematik
07/1996   Abitur am Bertolt-Brecht-Gymnasium in Dresden

Beruflicher Werdegang

seit 01/2005   wissenschaftliche Mitarbeiterin an der TU Darmstadt,
                   Fachbereich Mathematik, Arbeitsgruppe Stochastik
12/2002-12/2004   wissenschaftliche Mitarbeiterin am Forschungszentrum caesar
                   (center of advanced european studies and research), Bonn
11/1999-07/2001   wissenschaftliche Hilfskraft an der TU Dresden,
                   Fakultät Mathematik